## Science, Technology and Medicine in Modern History

General Editor: John V. Pickstone, Centre for the History of Science, Technology and Medicine, University of Manchester, England (www.man.ac.uk/CHSTM)

One purpose of historical writing is to illuminate the present. At the start of the third millennium, science, technology and medicine are enormously important, yet their development is little studied.

The reasons for this failure are as obvious as they are regrettable. Education in many countries, not least in Britain, draws deep divisions between the sciences and the humanities. Men and women who have been trained in science have too often been trained away from history, or from any sustained reflection on how societies work. Those educated in historical or social studies have usually learned so little of science that they remain thereafter suspicious, overawed, or both.

Such a diagnosis is by no means novel, nor is it particularly original to suggest that good historical studies of science may be peculiarly important for understanding our present. Indeed this series could be seen as extending research undertaken over the last half-century. But much of that work has treated science, technology and medicine separately; this series aims to draw them together, partly because the three activities have become ever more intertwined. This breadth of focus and the stress on the relationships of knowledge and practice are particularly appropriate in a series which will concentrate on modern history and on industrial societies. Furthermore, while much of the existing historical scholarship is on American topics, this series aims to be international, encouraging studies on European material. The intention is to present science, technology and medicine as aspects of modern culture, analysing their economic, social and political aspects, but not neglecting the expert content which tends to distance them from other aspects of history. The books will investigate the uses and consequences of technical knowledge, and how it was shaped within particular economic, social and political structures.

Such analyses should contribute to discussions of present dilemmas and to assessments of policy. 'Science' no longer appears to us as a triumphant agent of Enlightenment, breaking the shackles of tradition, enabling command over nature. But neither is it to be seen as merely oppressive and dangerous. Judgement requires information and careful analysis, just as intelligent policy-making requires a community of discourse between men and women trained in technical specialities and those who are not.

This series is intended to supply analysis and to stimulate debate. Opinions will vary between authors; we claim only that the books are based on searching historical study of topics which are important, not least because they cut across conventional academic boundaries. They should appeal not just to historians, nor just to scientists, engineers and doctors, but to all who share the view that science, technology and medicine are far too important to be left out of history.

*Titles include:*
Julie Anderson, Francis Neary and John V. Pickstone
SURGEONS, MANUFACTURERS AND PATIENTS
A Transatlantic History of Total Hip Replacement

Roberta E. Bivins
ACUPUNCTURE, EXPERTISE AND CROSS-CULTURAL MEDICINE

Roger Cooter
SURGERY AND SOCIETY IN PEACE AND WAR
Orthopaedics and the Organization of Modern Medicine, 1880–1948

Jean-Paul Gaudillière and Ilana Löwy (*editors*)
THE INVISIBLE INDUSTRIALIST
Manufacture and the Construction of Scientific Knowledge

Christoph Gradmann and Jonathan Simon (*editors*)
EVALUATING AND STANDARDIZING THERAPEUTIC AGENTS, 1890–1950

Alex Mold and Virginia Berridge
VOLUNTARY ACTION AND ILLEGAL DRUGS
Health and Society in Britain since the 1960s

Ayesha Nathoo
HEARTS EXPOSED
Transplants and the Media in 1960s Britain

Neil Pemberton and Michael Worboys
MAD DOGS AND ENGLISHMEN
Rabies in Britain, 1830–2000

Cay-Rüdiger Prüll, Andreas-Holger Maehle and Robert Francis Halliwell
A SHORT HISTORY OF THE DRUG RECEPTOR CONCEPT

Thomas Schlich
SURGERY, SCIENCE AND INDUSTRY
A Revolution in Fracture Care, 1950s–1990s

Eve Seguin (*editor*)
INFECTIOUS PROCESSES
Knowledge, Discourse and the Politics of Prions

Crosbie Smith and Jon Agar (*editors*)
MAKING SPACE FOR SCIENCE
Territorial Themes in the Shaping of Knowledge

Stephanie J. Snow
OPERATIONS WITHOUT PAIN
The Practice and Science of Anaesthesia in Victorian Britain

Carsten Timmermann and Julie Anderson (*editors*)
DEVICES AND DESIGNS
Medical Technologies in Historical Perspective

---

**Science, Technology and Medicine in Modern History**
**Series Standing Order ISBN 978–0–333–71492–8 hardcover**
**Series Standing Order ISBN 978–0–333–80340–0 paperback**
(*outside North America only*)

You can receive future titles in this series as they are published by placing a standing order. Please contact your bookseller or, in case of difficulty, write to us at the address below with your name and address, the title of the series and one of the ISBNs quoted above.

Customer Services Department, Macmillan Distribution Ltd, Houndmills, Basingstoke, Hampshire RG21 6XS, England

---

# Voluntary Action and Illegal Drugs

## Health and Society in Britain since the 1960s

Alex Mold
*Lecturer in History, London School of Hygiene and Tropical Medicine, UK*

Virginia Berridge
*Professor of History, London School of Hygiene and Tropical Medicine, UK*

First published 2010 by
PALGRAVE MACMILLAN

Palgrave Macmillan in the UK is an imprint of Macmillan Publishers Limited, registered in England, company number 785998, of Houndmills, Basingstoke, Hampshire RG21 6XS.

Palgrave Macmillan in the US is a division of St Martin's Press LLC, 175 Fifth Avenue, New York, NY 10010.

Palgrave Macmillan is the global academic imprint of the above companies and has companies and representatives throughout the world.

Palgrave® and Macmillan® are registered trademarks in the United States, the United Kingdom, Europe and other countries.

ISBN: 978–0–230–52140–7 hardback

This book is printed on paper suitable for recycling and made from fully managed and sustained forest sources. Logging, pulping and manufacturing processes are expected to conform to the environmental regulations of the country of origin.

A catalogue record for this book is available from the British Library.

Library of Congress Cataloging-in-Publication Data

Mold, Alex.
    Voluntary action and illegal drugs : health and society in Britain since the 1960s / Alex Mold and Virginia Berridge.
        p. cm.—(Science, technology and medicine in modern history)
    Summary: "Through a study of the voluntary activity around illegal drug use since the 1960s, this book explores wider issues in the changing relationship between the state and the individual in the making, provision and delivery of public services, and addresses the history of key issues in the development of contemporary health and social policy" – Provided by publisher.
        ISBN 978–0–230–52140–7
        1. Drug addiction – Treatment – Great Britain – History. 2. Drug addicts – Great Britain – History. 3. Voluntarism – Great Britain – History. I. Berridge, Virginia. II. Title.

HV5840.G7M67 2010
362.29'170941—dc22                                                2010002706

10  9  8  7  6  5  4  3  2  1
19  18  17  16  15  14  13  12  11  10

# Contents

# Contents

# Abbreviations

| | |
|---|---|
| APA | Association of Parents of Addicts/Association for the Prevention of Addiction |
| ARU | Addiction Research Unit |
| ASH | Action on Smoking and Health |
| CDCU | Central Drugs Coordination Unit |
| CDP | Community Drug Project |
| CFI | Central Funding Initiative |
| CND | Commission on Narcotic Drugs |
| DANGO | Database of Non-Governmental Organisations |
| DATs | Drug Action Teams |
| DDU | Drug Dependence Unit |
| DHSS | Department of Health and Social Security |
| DIG | Drug Dependency Improvement Group |
| DH | Department of Health |
| DL | DrugScope Library |
| DRG | Drug Reference Group |
| DTTO | Drug Treatment and Testing Order |
| EATA | European Association for the Treatment of Addiction |
| ECO-SOC | Economic and Social Council (of the United Nations) |
| ENCOD | European Network for Just and Effective Drug Policies |
| HO | Home Office |
| IHRA | International Harm Reduction Association |
| INPUD | International Network of People who Use Drugs |
| ISDD | Institute for the Study of Drug Dependence |
| JRF | Joseph Rowntree Foundation |
| LBA | London Boroughs Association |
| MDUG | Merseyside Drug Users Group |
| MH | Ministry of Health |
| NAT | National AIDS Trust |
| NCVO | National Council for Voluntary Organisations |
| NDDA | Nottingham Drug Dependents Anonymous |
| NDUDA | National Drug Users Development Agency |
| NGO | Non-Governmental Organisation |
| NHS | National Health Service |
| NICE | National Institute for Health and Clinical Excellence |
| NTA | National Treatment Agency |

| | |
|---|---|
| NTORS | National Treatment Outcome Research Study |
| NUAG | National Users Advisory Group |
| NUN | National Users Network |
| OUT | Oxfordshire User Team |
| SCODA | Standing Conference on Drug Abuse |
| TACADE | Teachers Advisory Council on Alcohol and Drug Education |
| TNA | The National Archives |
| QuADS | Quality in Alcohol and Drug Services |
| UKHRA | United Kingdom Harm Reduction Alliance |
| UN | United Nations |
| VCS | Voluntary and Community Sector |

# Preface

This book is based on research conducted as part of an Economic and Social Research Council funded projected entitled 'Drug User Patient Groups, 'User Groups' and Drug Policy, 1970s–2002', (grant number RES-000-23-0265) on which Virginia Berridge was the principal investigator and Alex Mold the researcher. We are grateful to the ESRC for this grant and also to the Department of Public Health and Policy, London School of Hygiene and Tropical Medicine (LSHTM) where this project was based. We also thank Sarah Mars, who played a key role in the initial establishment of the project, and this work built on her earlier Nuffield Foundation funded pilot project on drug user groups.

We would like to acknowledge the help and support of a number of people and institutions. Thanks go to archivists and staff at the Modern Records Centre, Warwick University, The National Archives, London and the Contemporary Medical Archives Centre at the Wellcome Library, London. We are particularly indebted to Professor Griffith Edwards who loaned us some of his personal papers, and Pauline Connor at the Department of Health archive in Nelson, Lancashire who helped facilitate access to recent material under the Freedom of Information Act. This book also draws heavily on oral history interviews, and we would like to thank all those individuals who gave up their time to speak to us. Libraries remain essential to historians despite advances in information technology, and we thank staff in the libraries of LSHTM, Senate House, the University of Birmingham, the Wellcome Library and the specialist drug research library until recently based at DrugScope in London.

Key sections of this book began life as conference and seminar papers, and we are grateful for the comments provided by audiences at the North American Conference on British Studies in Boston; the European Social Science History Conference in Amsterdam; the Society for the Social History of Medicine Conference at Warwick University; the Database of NGOs Conference at the University of Birmingham; the NGOs, Voluntarism and Health Workshop at the LSHTM; the Contemporary British History seminar at the Institute for Historical Research; the Centre for the History of Science, Technology and Medicine seminar at the University of Manchester; the Voluntary Action History Society seminar at the Institute for Historical Research; the History Seminar series at the London School of Hygiene and Tropical Medicine and a

British Academy conference on voluntarism. Other parts of this book have been published as articles in *Twentieth Century British History*, the *Journal of Policy History* and *Drugs: Education, Prevention and Policy*. We would like to thank the editors of these journals and the peer reviewers for their input.

Both authors have shared in the research for and writing of the book. Chapters 1, 2, 3, 4 and 7 were primarily the responsibility of Alex Mold and Chapters 5 and 6 of Virginia Berridge. Both worked on the Introduction and Conclusion, which were initially drafted by Alex Mold. Both authors have commented on each other's chapters and research has been jointly shared. We would also like to thank all those who contributed to this book through discussions and advice, particularly members of the Centre for History in Public Health at the LSHTM and Ingrid James for her administrative support. We are also grateful to the comments provided on the text by John Pickstone and Colin Rochester, by an anonymous peer reviewer on the proposal sent to Palgrave Macmillan, and to Michael Strang and Ruth Ireland who oversaw the publication of this book.

# Introduction

There is a great tradition of voluntary service in the United Kingdom and, in spite of all the rapid changes in our society, the impulse to serve is still strong and widespread.[1]

This statement, taken from the foreword to the 1969 Aves report on *The Voluntary Worker in the Social Services*, can be repeated today with the same degree of accuracy. According to the Home Office Citizenship Survey of 2003, some 17.9 million people in England and Wales had participated in formal voluntary activity over the last 12 months, devoting a total of 1.9 billion hours, a contribution worth at least £22.5 billion.[2] Voluntarism in Britain is clearly alive and well. Indeed, some figures would suggest that voluntary activity has increased over the last 40 years. Since the 1960s, there has been massive expansion in the number of voluntary organisations. In 1961 there were 1,182 registered charities, but by 2006 there were 168,600.[3] Moreover, in addition to formal charities many informal groups also existed. In 2004 it was estimated that there were as many as 250,000 'voluntary and community organisations' encompassing a whole host of bodies from Sunday league football teams to self-help groups.[4] The economic contribution of these and other organisations is always hard to measure, but in 2006 the National Council for Voluntary Organisations (NCVO) estimated that the voluntary sector as a whole had an income of £26.3 billion, and a paid workforce of at least 608,000 people.[5]

This numerical snapshot of contemporary voluntarism raises many questions. How can the significant increase in the size and scale of this kind of activity over the last four decades be explained? What does this increase mean, and does it matter? What is it that voluntary organisations actually do and how important has their contribution been?

1

Has their role changed over time, and if so, how? This book is dedicated to answering these, and other questions, by focusing on the specific case study of the voluntary activity that grew up around illegal drug use since the 1960s. As the above figures indicate, the voluntary sector is vast and somewhat amorphous: to make sense of it one area needs to be considered in detail. The drugs field is a particularly good example as illegal drug use was (and is) an issue that straddles three key areas of domestic policy, including health, society and criminal justice. Furthermore, drug use was one of a series of 'new' 'problems' discovered or, as in the case of poverty, re-discovered in the 1960s. The response to what the architect of the welfare state, William Beveridge, called the 'distressed minorities' (including groups such as the homeless, recent immigrants, unmarried mothers, homosexuals, ex-prisoners, drug users and alcoholics) came largely from organisations that existed away from the state.[6]

Of course, the provision of welfare services by non-state actors was nothing new, as a brief overview of voluntarism before the welfare state will indicate (see below). A significant role was played by voluntary organisations in the delivery of health services before 1948, and this did not entirely disappear even after the establishment of the National Health Service (NHS).[7] There was always something of a 'mixed economy' of welfare in the UK, but this became even more pronounced over time. The key players in this mixed economy were identified by the Wolfenden Committee on the Future of Voluntary Organisations in 1978. Wolfenden pointed to the existence of three formal systems in place to meet what it termed 'social need'. These were: the commercial system or the market; the statutory system; and the voluntary system.[8] In this book we will argue that since the late 1980s, these three systems have drawn closer together. While there was always some overlap between the market, the state and the voluntary sector, the degree of interpenetration between these has grown over the last 20 years. This can be observed, for example, in the introduction of markets into statutory welfare provision and in the appearance of 'hybrid' organisations such as 'social enterprises' that blend elements of all three sectors. It can also be seen in the way in which the state began to take on activities which were once the prerogative of the voluntary sector or of more loose and undefined clinical relationships. The rise of agencies such as NICE (National Institute for Health and Clinical Excellence) and, in the drugs field the NTA (National Treatment Agency) shows the state extending its role in this way. This was a process of 'rolling into the state' rather than the one of 'rolling back the state', a model which was often invoked.

The strengthening bond between voluntary organisations and the state also manifested itself in closer financial ties between the two sectors. In 2003–04 the NCVO estimated that 38 per cent of the voluntary sector's income came from the state, the largest single source of revenue.[9] Moreover, statutory sources of funding seemed to be the only stream of income that was increasing.[10] The apparent dependence of some voluntary organisations on the state for support has led some commentators to point to the demise of independent voluntarism. Frank Prochasaka, for example, contended that voluntary organisations in receipt of statutory funding risked becoming 'servants of the state'.[11] Yet there are three key reasons to believe that autonomous voluntarism persisted even as some organisations drew closer to the state. First, many of the voluntary organisations established in the 1960s (and even before, going back to at least the early twentieth century) were never entirely free of statutory involvement: they were tied to the state from the beginning, although the nature of those ties has clearly changed.[12] Second, there have been new forms of voluntary action taking place away from the state – this can be seen particularly in the emergence, since the late 1980s, of a drug user activist movement. Third, despite the drawing together of the market, the state and the voluntary sector, much remains distinctive about all three areas and the voluntary sector in particular. Government can be wary of too much interference as it is anxious to preserve what it sees as the unique value of voluntary organisations. As the Prime Minister, Gordon Brown, stated 'a successful modern democracy needs at its heart a thriving and diverse third sector. Government cannot and must not stifle or control the thousands of organisations and millions of people that make up this sector.'[13]

But at the same time, ideas about what the value of voluntary action constitutes, and the emphasis placed on this, have shifted over time. Indeed, while the 'impulse to serve' would still seem to prevail, the use to which it is put, and the context in which it is interpreted, has altered dramatically. By drawing on archival research, oral history interviews and document analysis, this book identifies three chronological phases in the history of voluntary action around drugs since the 1960s. These phases are also representative of key changes in voluntarism over the same period, and though these could be mapped onto a fairly standard narrative of change over time within voluntary action, this book also seeks to modify this story. Within each phase there were organisations and developments that seemed to run counter to overall trends. The first phase, from the mid-1960s to the late 1970s, is dealt with in Part I of the book. It saw the emergence of a large number of groups aiming to address different aspects of the drug 'problem', often in the belief that the welfare state could no longer comprehensively address social need. New social movements, which helped create new identities and lifestyles, also evolved new ways of dealing with new (and old) problems. At the same time, older approaches were also made use of. Self-help, for example, played a key part in the activities of many new groups. This section will assess the function of these old and new types of organisation, looking at their role as service providers, as campaigners and as repositories of self-help and mutual aid. The relationship between these groups and the state will also be considered, particularly around the central issues of funding and access to policy-making.

Part II examines the 1980s, the second key phase in voluntary action when voluntary organisations in the drugs field were called upon by the state in response to two crises: the explosion of heroin use and the arrival of HIV/AIDS. This was a move that recognised the value and expertise of these organisations, but was also indicative of broader shifts in health and social policy: chiefly the 'rolling back of the state' and the increasingly consumerist approach to the delivery and receipt of public services. However, as this section of the book will demonstrate, neither process was uncontested or driven solely by the state: voluntary groups fed back into these key developments. This interaction can also be observed in the third and final phase (dealt with in Part III), from the 1990s to the present, when voluntary groups drew closer to the market, in the form of 'business voluntary organisations' and 'social enterprises', but also closer to the state, in the form of 'service user groups'. This was not a linear process: there were countervailing and directly opposed trends, such as the emergence of independent drug user activist groups and the extension of state management through new agencies, like the NTA. Moreover, in the early years of the twenty-first century, voluntary and community groups have been tasked with enhancing 'civil society' and building up 'social capital', a development that recognises the importance of bodies outside the state in the development of strong, effective communities. This suggests that there is still a key role for voluntarism in British society. This book will chart and explain shifts in the nature of, and ideas about, voluntarism in this period through the prism of voluntary action around illegal drug use. But before the argument and content of this book can be outlined in detail, it is necessary to consider some of the language used to describe and define this area of activity.

## I.1 Definitions

Something of a 'terminological tangle' surrounds any discussion of voluntary action.[14] A large range of labels exist to describe seemingly similar kinds of activity, but each carries slightly different implications. 'Voluntarism' is used to refer to voluntary as opposed to compulsory activity, but it is also used to describe the philosophy behind voluntary action and the notion that this has a specific kind of value or importance. 'Voluntarism' is of course, not politically neutral, but a term with stronger political and social dimensions is that of 'activism'. To complicate matter still further, the sphere in which voluntarism and activism take place is also described in various ways. This book focuses on the space between the market and the state, but there are a number of alternative ways of labelling this arena and the groups that operate within it. One of the most commonly used terms is the 'voluntary sector', although this term is often associated with service providing organisations and government funding. The phrase 'voluntary and community sector' is frequently used in official documents, but another more recent favourite with the British government is the 'third sector', defined as 'non-governmental organisations which are value-driven and which principally reinvest their surpluses to further social, environmental or cultural objectives'.[15] But, as former civil servant and Professor of Social Policy Nicholas Deakin points out, using the term 'third sector' suggests a hierarchy with the 'third sector' placed beneath both the state and the market.[16] Furthermore, 'third sector' implies that the three sectors are distinct and self-contained and, as this book will demonstrate, this was rarely the case. Another term sometimes used is the 'independent sector', but it is not clear what the independent sector is supposed to be independent of, or how this supposed independence could be judged. In fact, as Stein Kuhnle and Per Selle point out, 'independent' organisations are rarely independent of either the state or the market.[17]

Yet more labels have been applied to the specific organisations that operate within the space between the state and the market. The term 'charity', of course, has an ancient history, and in the UK, it has specific legal meaning as organisations can apply for charitable status.[18] This has a number of benefits, such as exemption from certain kinds of tax, but also some restrictions, such as limitations to political campaign work. For this reason, by no means all voluntary organisations were or are charities. In American studies the term 'non-profit' is often used.[19] But, many modern organisations do make a profit, even if this profit is re-invested rather than divided up between shareholders. Another popular label, particularly in the context of international development, is 'non-governmental organisation' or NGO. This term has most frequently been used to refer to groups working in the third world or the global 'South', but has recently been taken on by some historians to describe UK-based organisations.[20] The term NGO is useful as it draws attention to the wider role that such organisations play, particularly around socio-political action. Yet, there are also problems with the label NGO. Some groups, such as Narcotics Anonymous (NA), reject the notion that they are political 'players' altogether, although they clearly have a socio-political function. Other organisations, through their complete dependence on the state for financial support, are questionably 'non-governmental' and yet would still see themselves, and be seen by others, as distinct from the state. Indeed, 'NGO' was not a label used by British organisations to describe themselves until very recently, and even then it is mostly taken on by organisations that work internationally.

Despite the plethora of labels available to them, most groups continue to refer to themselves as 'voluntary organisations', although some of the more politically active prefer the label 'activist'. This self-identification is one of the key reasons why we decided to use the term 'voluntary organisation' in this book. Throughout the text the 'voluntary sector' is used to describe a whole group of organisations (with 'non-statutory sector' sometimes used to avoid endless repetition); 'voluntary organisation' to describe individual groups; and 'voluntarism' to describe voluntary action in general. We will also use other terms where appropriate to describe different, but related concepts such as 'social movement', 'social capital' and 'civil society'.[21] To define what constitutes a 'voluntary organisation', the structural operational definition devised by Lester Salamon and Helmut Anheier at Johns Hopkins University in the 1990s will be used. Although this approach was developed in the US, (a country with a different voluntary tradition to the UK) the structural operational definition was intended to be used when comparing voluntary action cross-nationally, thus giving the model wider applicability. Salamon and Anheier identified five key features of voluntary organisations: first, these should be formal organisations governed by a set of rules; second, they should be institutionally separate from government; third, they should be non-profit distributing and primarily non-business; fourth they should be self-governing; and finally, they should contain a meaningful degree of voluntary action, either through philanthropy or through voluntary citizen involvement.[22]

Of course this model is not without its flaws. A particularly problematic area is defining what constitutes 'a meaningful degree of voluntary action.' Voluntary organisations may rely on the efforts of unpaid volunteers, but many also employ paid staff, and some have no volunteers working for them at all. Volunteering, as Jos Sheard has pointed out, is not necessarily synonymous with the voluntary sector.[23] A distinction can be made between what David Billis calls 'pure' and 'unambiguous' organisations (the groups which are made up entirely of volunteers) and those that are more 'bureaucratic' and 'ambiguous' (the organisations which employ paid staff).[24] Furthermore, there are organisations that are in transition between these different categories, and many of the barriers between seemingly fixed entities are blurred. For example, a volunteer within a particular organisation might also be a client using its services and/or a member of its management board.

Yet, despite these differences and difficulties 'voluntary' remains a useful term. Not only is it the label organisations used (and continue to use) to describe themselves, but it also communicates something about the nature of this activity. 'Voluntary' points to the independent origins of these groups and also highlights the fact that these organisations were often created with a sense of a moral, social or political purpose.[25] Indeed, using the label 'voluntary' also enables us to link voluntary action in contemporary history with voluntary action in the more distant past. Placing voluntary organisations in their broader historical context allows for a deeper exploration of their specific origins, motivations and activities. At the same time, historical context is also important for the more general picture, because, as a brief consideration of the history of voluntary action before the 1960s indicates, there are important continuities as well as changes within voluntarism in the past, and in the present.

## I.2   Voluntary action before the 1960s

In his 1948 report *Voluntary Action* Beveridge separated voluntary activity into two types: philanthropy and mutual aid.[26] Yet this distinction between 'other regarding' and 'self regarding' forms of voluntarism was well established even before the 1940s, with both philanthropy and mutual aid playing a key part in British welfare provision since at least the medieval period.[27] Before the nineteenth-century, philanthropy chiefly manifested itself in charity provided by the church or other religious groups.[28] At the same time, mutual aid and self-help could be found in medieval fraternities and religious guilds.[29] Yet it was during

the nineteenth century that voluntary action became particularly preva-
lent. A rapid increase in the number of charitable organisations during
this period led some commentators to see the nineteenth century as
the 'golden age' of the voluntary association.[30] In the 1880s *The Times*
claimed that the income of London charities exceeded that of several
nation states including Sweden, Denmark and Portugal, and by the
1890s, according to Prochaska, the average middle class household was
spending more on charity than it was on any other item except food.[31]
Some historians have seen the growth of Victorian philanthropy as an
example of increasing social control of working class lives by the middle
and upper classes. More recently, it has been argued that philanthropy
can instead be regarded as forming the basis of a consensus within
which the middle and working classes could be brought together to
create a 'peaceable kingdom' which excluded the 'undeserving' poor.[32]
Furthermore, the continued existence of mutual aid and self-help alongside philanthropy pointed to the significance of 'self regarding' as well
as 'other regarding' forms of voluntary action. Particularly important
were the friendly societies. These took a number of different forms, but
by far the most common was to provide insurance against sickness and
payment of members' funeral costs. Run by the members themselves,
the friendly societies had a larger membership than the trade unions.[33]

   Voluntarism of both the philanthropic and mutual aid type played
a significant role in the funding and delivery of healthcare before the
establishment of the National Health Service. Hospitals in this period
were either run by the local authority, or through the poor law and
public assistance, or in the voluntary sector.[34] By the late 1930s voluntary hospitals provided the largest proportion (41 per cent) of all beds.[35]
Voluntary hospitals were funded through a mixture of charitable donations by wealthy patrons and through worker contributory schemes.[36]
This amounted to a significant role for voluntary groups in the provision
of health services before the establishment of the NHS, and the drug
and alcohol field was no exception.[37] In fact voluntarism and private
provision was even more important in this area because of the failure
before the First World War to establish a national rate funded treatment
system. The intention was that the inebriates acts would provide such a
system, but this aim was never achieved.[38] Particularly important were
Christian organisations such as Spelthorne St Mary, where the Sisters of
the Community of St Mary the Virgin treated female alcoholics from
1879 onwards.[39] For the more affluent alcoholic, there were also private
institutions. Voluntary services were, therefore, an important part of
providing services for drug and alcohol users before the welfare state.

Indeed, an array of voluntary action in all areas continued to grow even as the state came to play an increasing part in the provision of welfare over the twentieth century. As assumptions about the causes of poverty shifted, the state became more involved in social aid.[40] This enhanced role for the state led to a re-examination of the 'moving frontier' between voluntary and statutory activity. For Sidney and Beatrice Webb, writing in the 1910s, there had been a shift from the notion of voluntary and statutory sectors operating in 'separate spheres' to something close to a partnership between the two actors by the early years of the twentieth century. To some extent, as Jane Lewis points out, this was a misconception as the state and the voluntary sector had worked together in the past. Rather, she suggests that during the first half of the twentieth century there was a renegotiation of this partnership in the wake of increased state provision. The voluntary and the statutory sectors drew closer together; the state even began funding voluntary organisations in some cases.[41] Geoffrey Finlayson has also pointed to the closeness of the relationship between voluntary organisations and the state in this period. He maintains that this bond became stronger still during the 1930s, a move which suited both the state and voluntary organisations. For the state, using charitable organisations was often cheaper than statutory provision, and for voluntary organisations, the state was able to provide the more costly and greater range of services being demanded by groups and individuals. Finlayson asserts that though the frontiers of the state expanded in this period, this was not to the extent that voluntary action was excluded.[42]

It is, however, an inescapable fact that the statutory provision of welfare increased dramatically in the first half of the twentieth century. The legislation of the late 1940s, establishing universal free secondary education, a single co-ordinated system of national insurance, unemployment benefits and free healthcare culminating in the introduction of the National Health Service in 1948 was credited with establishing a 'welfare state' that would provide adequate social services for all. The reasons for the growth of the welfare state are complex and multi-faceted and cannot be explained adequately here, instead it is more pertinent to look at the impact its establishment had on the voluntary sector.[43] Many contemporaries felt that the introduction of statutory welfare provision would lead to the 'withering away' of the voluntary sector, its work becoming superfluous in the shadow of a comprehensive welfare state.[44] Some sections of voluntary activity, such as the voluntary hospitals, did become incorporated within the welfare state, but voluntarism as a whole did not disappear.[45] Indeed, there were signs even from the late 1940s that this was unlikely to occur. Beveridge's third report on welfare, *Voluntary Action*, published in 1948, set out a clear and enduring role for voluntary organisations.[46] The existence of the voluntary sector was not threatened by the welfare state, and went through a period of consolidation in the 1950s. A review of voluntarism in 1955 found it to be particularly strong in three areas: maternity and child welfare, mental health and services for the blind.[47] Yet, as outlined by the Younghusband report in 1959, voluntary action in this period was usually seen as largely supplementary to statutory provision.[48] The voluntary sector was clearly the 'junior partner in the welfare firm.'[49]

## I.3   Voluntary action and illegal drugs since the 1960s

By the 1960s the relative position of the voluntary and statutory sectors altered somewhat as needs, and the perception of these, began to change. A series of dramatic exposures during this period highlighted significant deficiencies in welfare provision in a number of areas, prompting the establishment of organisations to campaign for, and provide, improvements.[50] As Finlayson noted 'it was increasingly recognised that the state could not cope alone. There was a need for active citizens *inside* the state but also *outside* it.'[51]

The drugs field was one area where gaps in existing provision were to be found. Until the 1960s illegal drug use in Britain was comparatively rare and attracted little interest or concern. Opiate addiction was largely confined to a small, mostly middle aged, middle class population who had become addicted to drugs following treatment for another condition. There was also some cannabis use amongst the West Indian community during the 1950s, but this practice did not spread to the wider counter-culture until the mid- to late 1960s. In 1965 there were 626 convictions for the possession of cannabis, but by 1967 there were 2,393.[52] At the same time, the use of other kinds of drugs also appeared to be increasing. In 1959 there were just 454 drug addicts known to the Home Office, but by 1969 there were 1,462: small numbers in comparison today, but a significant increase nonetheless.[53] Moreover, drug users were younger – in 1959, 11 per cent of reported addicts were under 35 years of age, by 1964, 40 per cent were in this group – and more likely to have started taking drugs recreationally rather than as a result of medical treatment for another condition. It is estimated that 94 per cent of newly reported addicts in 1964 were of non-therapeutic origin.[54]

This rising tide of illegal drug use seemed to present a number of problems that were dealt with by the state through a series of legal and medical measures. The Dangerous Drugs Act was revised three times over the 1960s to increase penalties for the possession of drugs such as cannabis and to provide for the control of the prescription of drugs such as heroin.[55] Governmental advisory committees were convened to examine heroin use in 1959 and again in 1964; cannabis and LSD in 1968; and police powers of search and arrest in 1969.[56] But the drug that provoked the most concern was heroin. Following the report of the *Second Interdepartmental Committee on Heroin Addiction*, published in 1965, specialised Drug Dependence Units (DDUs) were set up to treat heroin addiction in 1968. Based in London teaching hospitals, the DDUs were intended to both treat the addict and control the spread of addiction.[57] However, some groups and individuals felt that the DDUs only addressed one aspect of the growing drug 'problem' and voluntary organisations were also established to provide legal services to drug users, to offer social support and practical advice, to campaign on their behalf, to collate information about drug use and to co-ordinate the sector as a whole.

Yet this belief that drug use was not adequately being dealt with by the state was not just rooted in the apparent failings of statutory welfare, but also in a supposedly 'new' form of politics and political activity. The appearance of 'new social movements', such as those concerned with civil rights, women's rights and the environment, drew attention to previously marginalised groups and interests.[58] Many new voluntary organisations that came into being during this period were established around these issues. Some were orientated towards service provision, others took on a more campaigning role, and many combined both. This pattern of overlapping origins and functions could also be found in the drugs field. The charity Release, for example, which was established in 1967 to provide legal assistance to people arrested for drug offences, also undertook campaigning work in a number of areas.[59] Release saw their defence of the legal rights of the drug user as a way of providing individual aid, but also as a way of critiquing government policy on drugs. They argued that the drug problem could not be solved by the 'conventional means of criminal reprimand'; they felt that 'medical or social solutions [were] more likely to be successful.'[60]

Despite their challenging position on drug policy Release, and many other organisations working in the drugs field, received funding from the state. This was in line with a more general increase in statutory funding of the voluntary sector. Central government grants to voluntary organisations in 1970–71 were estimated to be in the region of £2.5 million; by 1976–77 this had risen to £35.4 million.[61] The influx of government money to the voluntary sector did raise fears about the dependence of voluntary organisations on the state for financial support, but as Maria Brenton points out, 'Dependency does not rest all on one side, however, for this financial relationship also disguises the extent of increased dependence of government on the voluntary sector.'[62] Indeed, as the first three chapters in this book will show, voluntary organisations working in the drugs field that were in receipt of government funds were afforded considerable freedom of action. Government attempts to interfere in the activity of voluntary groups during the 1960s and 1970s were comparatively rare. A detailed consideration of how this relationship worked in practice in the drugs field will demonstrate that the 'partnership' between the voluntary and statutory sectors was finely balanced, with both partners losing and gaining by becoming more closely entwined.

This relationship between the state and the voluntary sector shifted again during the 1980s. In this period the New Right adopted the idea that voluntary organisations could contribute something of distinctive value to welfare provision. The Conservative government led by Margaret Thatcher regarded the state as an inefficient and ineffective provider of welfare, and considered its monopoly on the provision of services to have resulted in a culture of passivity and dependence amongst welfare recipients.[63] The suggested solution to this problem was to 'roll back the state'; to reduce the role of central government in the provision of welfare. The 'rolling back of the state' was to be achieved in two closely related ways. First, by placing greater emphasis on the involvement of voluntary organisations in the delivery of health and social services; and second by creating a 'market' in welfare, allowing statutory and non-statutory bodies to bid for contracts to provide specific services.[64] In both these developments the role of the voluntary sector was crucial; not only was it seen as more responsive, more innovative and more cost-effective than the statutory sector, but using the voluntary organisations was also to reduce reliance on the state through the 'invigorating' experience of self-help and community care.[65]

The drugs field was an important test area for this general policy. During the 1980s illegal drug use in Britain appeared to be increasing at an alarming rate and spreading across the country on an unprecedented scale. An apparent growth in the use of heroin caused particular concern: the number of known heroin addicts rose from just over 2,000 in 1977 to more than 10,000 by 1987.[66] Moreover, heroin use was being reported in urban areas throughout the country.[67] This was in contrast to previous decades, when it was thought that drug use was largely confined to London.[68] To combat this seemingly worsening problem, the government introduced the Central Funding Initiative (CFI) for drug services in 1982. Under the initiative a total of £17.5 million pounds was awarded to organisations providing services to drug users throughout the country between 1983 and 1989. Crucially, the CFI was open to service providers in both the statutory and voluntary sector. Of the 188 grants issued under the CFI, 58 per cent went to statutory organisations and 42 per cent to non-statutory groups.[69] This significant level of support for voluntary organisations was driven by the recognition that statutory welfare services in the drugs field were inadequate, but also by a much broader strategy for involving the non-statutory sector in health and social service provision. Yet, as Chapter 5 of this book will reveal, this did not amount to a retreat by the state: central government retained a critical role in initiating and directing the actions of voluntary groups.[70]

A similar pattern can be observed in the way both voluntary organisations and the state responded to another crisis in the drugs field: the discovery of HIV/AIDS amongst injecting drug users. At a time when infectious diseases had no longer seemed a problem, the appearance of a new and rapidly fatal syndrome was a major shock to health and welfare authorities and to the gay communities on whom it first impacted. The new syndrome arrived in a vacuum: there were no policy traditions and no AIDS related voluntary or statutory organisations to respond to it. The initial reaction was a reversion to earlier traditions of 'pure' voluntary activity. Gay men set up organisations with no statutory funding, providing the service infrastructure which would otherwise have been lacking. This voluntarism also linked with the drugs field: in part because of the connection which HIV/AIDS brought between the two worlds, in part because of existing personal connections. Gay activism stimulated a new user activism in the drugs field as well. Policy change also helped bring this about. The emphasis within drug policy changed in the wake of AIDS to focus on attracting the user into services, so the role of the user became more visible. AIDS and its elevation of the role of the user, either the gay user of treatment and participant in trials, or the drug user accessing needles and methadone, was a forerunner of the more general patient consumerism and involvement which became a mantra for political parties in the 1990s.

The response to AIDS encapsulated classic voluntary/ state dilemmas: the new organisations feared incorporation by the state. But they also needed state funding in order to operate, to reach out to their constituency, and to gain access to policy-making. AIDS also threw up new dilemmas for government. The gay run voluntary sector had to be controlled – but not overtly. A new form of state funded voluntary organisation, the National AIDS Trust, was a precursor of later organisations in the 1990s and the early twenty-first century. There was a process of 'rolling into the state' as well as 'rolling out of it'. This dance between voluntarism and the state in the 1980s pre-figured the later role of the National Treatment Agency in the early twenty-first century. That Agency was set up as a Special Health Authority, not as an ostensible voluntary organisation and this organisational change underlined the enhanced overt role for the state in the early part of the new century.

The further impact of markets and a more consumer orientated approach to the delivery of public services can also be observed in the fate of voluntary organisations since the 1990s. The introduction of the internal market within the NHS, following the publication of the White Paper, *Working for Patients* in 1989, led to the introduction of local authority or GP 'purchasers' and statutory, voluntary or private 'providers' of services.[71] This more marketised approach to public services resulted in the proliferation of different service providing groups, but many of these were often tied (to a greater or lesser extent) to the state. The introduction of service agreements, or contracts, between a local authority purchaser and voluntary or private sector providers in the 1990s had a significant effect on the way voluntary organisations operated. Contracts imposed professional standards of assessment, management and evaluation on voluntary agencies. For many volunteers, this appeared to threaten the very nature of voluntarism. Some organisations were also concerned that contracting could compromise their campaigning roles and diminish their autonomy. Still others were worried that contracting would make their organisations more bureaucratic and formalised.[72]

Evidence from the drugs field suggests that at least some of these fears have been realised. Statutory support for voluntary organisations in the drugs field continued apace in the 1990s, initially spurred on by the need to combat HIV/AIDS, but more recently focusing on treatment provision and on breaking the supposed link between drug use and crime. Key organisations, like Addaction and Turning Point, have become very big service providers and operate as 'social businesses'. Questions have been raised about the ability of these organisations to remain independent of government as they rely heavily on contracts with statutory authorities to provide services. At the same time, the creation of a quasi-market in public service provision has also resulted in greater attention being paid to the views of service users themselves. Since the 1990s a string of schemes have tried to involve the patient or service user in decisions about their own care and wider service provision. This has resulted in the establishment of service user groups which offer input into treatment and social services at a local and national level. However, doubts remain about the power of these groups to effect real change.

Indeed, some of the sharpest critics of these service user groups have been other drug users. In addition to *service* user groups there are *activist* user groups, groups of drug users that are not usually directly connected to a statutory service and who offer an independent and challenging view on existing policy and provision. The appearance of these groups suggests a persistence of independent voluntarism around drug use, despite the drawing together of the state, market and voluntary sector. Indeed, although the three sectors have moved closer together, there is much that remains distinctive about voluntary organisations. Recent government interest in the concept of social capital – the notion that social networks have value – may yet help to preserve an independent role for the voluntary sector. This book will consider this, and other key developments, as it surveys voluntarism from the 1960s to the present.

# Part I
## 1960s–1970s

# 1
# The 'Old': Self-Help, Phoenix House and the Rehabilitation of Drug Users

In December 1969 the Phoenix House therapeutic community for the rehabilitation of drug users welcomed its first group of residents. An information leaflet about life in the community began 'Phoenix House is a programme for beating drug addiction...We are not a hospital. We are a community. We are staffed mostly by ex-addicts. We tell you how to make it. More important, we show you how to make it.' There were, the leaflet explained, 'only two rules, no drugs and no physical violence. People think the hardest step is giving up drugs. It isn't. The hardest step is deciding you'll try to do something for yourself.'[1] Phoenix House aimed to get addicts to confront and then address the reasons behind their drug use, a process which was designed to 'rehabilitate' the individual. 'Rehabilitation', according to the governmental Advisory Committee on Drug Dependence (ACDD) aimed to 're-educate the individual to live without drugs and to assume or resume a normal life.'[2] Rehabilitation was regarded by the ACDD as the second stage of a treatment process that began with the addict withdrawing from drugs, something that could be achieved through the treatment offered in the NHS Drug Dependence Units (DDUs).[3] The intention was that addicts would come off drugs through treatment at the DDU, but in order to stay off they should undergo a programme of rehabilitation, something that was almost exclusively provided by organisations like Phoenix House that operated within the voluntary sector.

This reliance on voluntary organisations can partly be related to the long tradition of voluntary work in the rehabilitation of alcoholics. Inebriate hospitals run by Christian groups provided residential treatment for alcoholics from the late nineteenth century onwards.[4]

But Phoenix House and organisations like it were the product of more than the continuation of particular modes of treatment within the drug and alcohol field. A number of different strands fed into the making of Phoenix House, and by exploring these in detail this chapter aims to shed light on broader factors at work within voluntarism during the 1960s and 1970s. Phoenix House can be seen as the manifestation of a very old form of voluntary action: self-help. The explicit appeal to the drug user to 'do something for yourself' was indicative not just of the idea that addiction was an individual problem that could only be overcome by the user, but was also a reflection of a long-running tradition of self-help and mutual aid that had been part of voluntary action in Britain for centuries. At the same time, new ideas about the self and how the self could be improved came into being in the 1960s and '70s. American models, such as Alcoholics Anonymous and the therapeutic communities for dealing with addiction that sprang out of this, were particularly influential within the drugs field. More broadly, this period also saw the growth of a large number of mutual-aid and self-help orientated organisations across a number of different fields, from playgroups to support for single mothers.[5] Some of these new groups took a more activist stance than similar organisations of the past, acting as pressure groups pushing for changes in the direction of scientific research and service provision. Phoenix House must be set against such changes, but is also an example that allows us to interrogate these in detail.

This chapter begins by considering self-help during the 1960s and 1970s. It will examine some of the reasons for an apparent increase in self-help, by exploring the social, cultural and political context that facilitated the development of this kind of activity. The chapter will then go on to consider the establishment of the Phoenix House therapeutic community for the rehabilitation of drug users, suggesting that it drew on two different models of the 'therapeutic community'; the 'democratic', which originated in Britain during the 1940s, and the 'concept-based', which had developed in America in the late 1950s and early 1960s. Here was an example of 'policy transfer', with ideas being imported from another national context into British social policy by voluntary organisations. But, blending these two styles of therapeutic community was not without its problems, as will be seen in the third section of this chapter. Democratic therapeutic communities placed a greater emphasis on the role of professionals when compared to concept-based communities, a tension that was mirrored in conflict between the health professionals involved in Phoenix House and the ex-addict staff. In the final section of this chapter an assessment of the broader significance of this conflict, and the work that Phoenix House did, will be made. By trying to help drug users do something for themselves Phoenix House did create a role for the user in his or her treatment and the treatment of others. In this way, 'old' notions of self-help and mutual aid that had been central to voluntary action for centuries helped to create a 'new' role for the drug user. Old forms of voluntarism were thus remade and revitalised to deal with new problems in the 1960s and 1970s.

## 1.1   The self and self-help in the 1960s and 1970s

The term 'self-help' is often associated with Victorian values and particularly Samuel Smiles' treatise *Self-Help*, first published in 1859.[6] Yet, as Deakin has noted, Smiles' notion of self-help was not just individualistic: self-help could be collective as well as centred on the individual.[7] A similar understanding can be found in more recent expositions on self-help. Frank Riessman, a social psychologist and self-help group advocate, and David Carroll, a writer interested in consumer affairs, noted that, 'For many years we have struggled with the term *self-help* [their italics]. Considering that most of our work relates to mutual aid groups, it has always seemed peculiar to call this activity *self*-help when much of what happens is group behaviour.' For them, mutual aid was a part of self-help, but this could operate at various levels: with the individual, the group and wider society. What these different forms of self-help had in common, according to Riessman and Carroll, was 'an emphasis on promoting latent inner strengths, and the special understanding that comes from proximity to the problem or need.'[8]

In the 1960s and 1970s a series of problems were identified as requiring self-help or mutual aid. Some were supposedly new problems like drug taking; others were old problems like poverty.[9] So, while self-help had always been an important feature of voluntarism in Britain, this seemed to increase during these decades. Indeed, some commentators saw the 1970s as 'the self-help decade', as an 'era of self-determination' within which self-help was assuming a greater level of importance than it had done in the past.[10] This new self-help activity took on a variety of forms; one particularly influential development was the emergence of the organised consumer movement in the late 1950s and early 1960s. Although the demands of groups such as the Consumers' Association (established in 1956) were not necessarily explicitly articulated around the notion of self-help, their work could be partially understood in this light. Organisations like the Consumers' Association were run by consumers themselves to press for the production of better goods and services, and later on, consumer 'rights.'[11] These ideas and the development of a wider 'consumer society' diffused into other fields. This can be seen in health, and though the precise relationship between consumerism and the creation of patient groups requires further investigation, it is clear that there was contemporary interest in the idea of the patient as consumer.[12] The rise of health consumerism has been used to explain the significant increase in the number of patient groups established in this period. Bruce Wood found that 66 disease-related patient associations were founded between 1960 and 1979, compared with the 14 established between 1940 and 1959.[13] One of the fastest growing areas was around self-help for individuals suffering from specific diseases, with groups such as the Brittle Bone Society (1971), the Friends of Asthma Research Council (1972), the Spinal Injuries Association (1974) and the National Eczema Society (1975), all being founded at this time.[14]

Although such growth could be related to a corresponding increase in the number of diseases and conditions identified by modern medicine, this does not explain how or why the nature of these groups appeared to be different to many of the voluntary organisations that had operated around health in the past. The organisations founded in this period tended to be led by patients themselves and not, as they had been in previously, run by health professionals.[15] This shift from doctor-led to patient-led groups is significant, particularly as this coincided with an apparent decline in professional authority. Harold Perkin contended that there was a 'backlash against professional society' from the 1970s onwards as politicians and other commentators questioned the power and privilege of professionals of all kinds, including Civil Servants, university academics, teachers and those in the 'caring professions.'[16] Doctors and the medical profession came under particular attack. A number of theorists such as Ivan Illich, Michel Foucault and Thomas Szasz provided an intellectual critique of medical power and the way it denied agency to individuals.[17]

One area of medicine that came in for special criticism was psychiatry.[18] According to the sociologist Nick Crossley, scandals around the treatment of mental illness in Britain during the 1950s and the appearance of anti-psychiatry in the 1960s created a 'field of contention' – a climate within which the consensus on mental health could be challenged.[19] The work of 'anti-psychiatrists', such as R.D. Laing and David Cooper, specifically questioned the notion of 'madness' and conventional treatment for psychiatric 'illness.'[20]

Although anti-psychiatry was not specifically a patient led protest, some of these ideas clearly influenced patient groups and the emerging mental health users' movement.[21] Within this movement, notions of self-help combined with an intellectual criticism of psychiatry. Individuals who had personal experience of mental illness began to form groups and organisations attacking conventional psychiatric practice and often attempted to offer another way of dealing with mental illness in its place. People Not Psychiatry, established in 1969, provided a network of support for those experiencing mental distress as an alternative to hospitalisation.[22] The Mental Patients' Union (MPU), created in 1971, also demanded the establishment of alternatives to psychiatric treatment. Crucially, the MPU wanted these new forms of treatment to be under the control of patients themselves. This emphasis on self-help was partly a legacy of many early members' experiences in Paddington Day Hospital, a democratic therapeutic community.[23] These ideas seem to have had a wider impact as other mental health organisations began to change their approach. For example, in the 1970s the previously somewhat paternalistic National Association for Mental Health (NAMH) changed its name to MIND and moved towards becoming a civil rights organisation, campaigning for the rights of mental health users.[24]

At the same time as self-help activity seemed to be growing in areas such as mental health, fundamental questions were also being asked about the nature of the 'self' that required helping. Nikolas Rose has contended that the notion of the 'self' was constructed during the latter half of the nineteenth-century by the so-called 'psy' disciplines: psychiatry and psychology.[25] Yet, as more recent work by Mathew Thomson has demonstrated, this self was not a fixed entity, undergoing considerable changes throughout the twentieth century. In the 1960s and 1970s the self was often spoken of in terms of 'growth' and 'discovery' and various therapeutic techniques were developed to allow individuals to uncover and develop their true selves. These included encounter groups (which will be discussed in greater detail later), Gestalt therapy, bio-energetics, psychodrama and yoga.[26] One method of self-exploration particularly popular within the counter-culture was to use illicit drugs, especially psychedelics such as LSD, magic mushrooms and cannabis.[27] But, there was also a degree of recognition by individuals both inside and outside the counter-culture that drugs could do harm to the self, and a strand of self-help began to develop around assisting people to overcome their problems with drugs. New ideas about the self and self-help, and ways of dealing with the challenges to the self posed by drug use, were thus able to take root within a receptive environment.

A particularly influential model was Alcoholics Anonymous (AA) and the organisations that developed out of this. Self-help and mutual aid were at the heart of AA, founded in 1935 in Akron, Ohio by two former alcoholics, stock-broker Bill Wilson and physician Robert Smith. 'Bill W' and 'Dr Bob', as they became known in AA circles, found that their sharing of experiences helped them to remain abstinent from alcohol.[28] The AA self-help philosophy spread to the UK and around the world, so that by 1950 world-wide membership was estimated to be over 100,000.[29] Because AA did not tolerate the presence of drug users, a group of heroin addicts established Narcotics Anonymous in California in 1953. NA was slower to spread; the first NA group in Britain did not appear until 1979.[30]

A more immediately successful British offshoot of the AA/NA self-help approach to drug and alcohol addiction was what became known as the 'concept-based' therapeutic community. The 'concept' was the belief that drug addiction was indicative of deeper problems that could be addressed through group therapy and the imposition of a strict hierarchy.[31] This approach originated in Synanon, a self-help group and later a residential community, established by former AA member Chuck Deiderich in California in 1958.[32] Deiderich and a group of other ex-drug users had found themselves unwelcome at their local AA meeting, so they established their own group drawing on some of AAs ideas, but also developing new approaches. What was particularly significant about Synanon was its use of ex-addicts as therapists. This model was emulated by other concept-based therapeutic communities for drug users set up in the 1960s, such as Daytop Village and Phoenix House.[33] Phoenix House was established in New York in 1966 by psychiatrist Efren Ramirez and run by ex-addicts and ex-Synanon residents such as Frank Natale, and professionals including the psychiatrist Mitchell Rosenthal.[34] As Rosenthal pointed out, the involvement of professionals did not necessarily win the programme acceptability. He commented that in America 'the TCs [therapeutic communities] operated far from the medical and mental health mainstreams.'[35] Moreover, the presence of psychiatrists did not diminish Phoenix House's emphasis on self-help and mutual aid: 'Self-help, we are fond of saying, is the mechanism for change.'[36]

Self-help was also central to other, older models of the therapeutic community. Therapeutic communities in the UK predated the establishment of Synanon and Phoenix House in the US, but they took a rather different form. According to the sociologist Nick Manning, therapeutic communities can be characterised as 'residential groups set up to pursue therapeutic or educational change in a manner at odds with conventional psychiatric treatment, particularly in the almost exclusive use of group experiences as the medium of therapy.'[37] This broad definition encompasses a wide range of practices, and different types of therapeutic community developed in different places at different times. Some commentators locate the origins of the therapeutic community as far back as the late eighteenth century, drawing parallels between the central tenets of 'moral treatment' – minimal restraint and engaging the patient in useful work – and the ethos of more modern systems.[38]

However, the actual term 'therapeutic community' was first used in Britain by Thomas Main in 1946 to refer to experimental work carried out during, and immediately after, the Second World War at the Northfield Military Psychiatric Hospital in Birmingham.[39] Together with other psychiatrists such as Bion and Foulkes, Main created a system of group therapy where the community was 'both patient and the instrument of treatment.'[40] Within the community less emphasis was placed on social hierarchies and on the distinction between doctor and patient, an idea further developed by Maxwell Jones at the Belmont Hospital, Surrey. This became known as the 'democratic' type of therapeutic community, a model that contrasted with the emphasis placed on hierarchy in many of the American therapeutic communities.[41]

As will be seen below, the creation of therapeutic communities to deal with drug addiction in the UK tended to draw on both of these traditions, but combining the two styles of therapeutic community was often problematic.

## 1.2 The establishment of Phoenix House

The immediate motivation for the establishment of Phoenix House and similar therapeutic communities in the late 1960s was the perceived need to provide for the rehabilitation of drug addicts. As discussed in the Introduction, rising drug use, and the use of heroin in particular, caused concern in this period, leading to a number of different statutory responses at the legal, medical and administrative level.[42] The main medical response was to encourage addicts into treatment in the newly created Drug Dependence Units, which were designed to both treat heroin addicts and control the spread of the 'social disease' of heroin addiction. To achieve this, the DDUs offered two kinds of treatment: maintenance, where an addict would be prescribed a dose of an opiate drug in order to prevent him or her from buying drugs on the black market; and withdrawal, where an addict would be removed from drugs over a short period of time.[43] Yet, withdrawal was not seen as the end of the treatment process: addicts also needed to be 'rehabilitated.' Although the ACDD asserted that 'rehabilitation begins with the first contact of the addict with the out-patient clinic [the DDU]' they believed that in-patient facilities were best equipped to rehabilitate the addict. The committee offered no particular view as to what these facilities should comprise, noting that 'there is room for much experimentation and it would be unrealistic to advocate any particular method or methods.'[44] Moreover, the ACDD were clear that rehabilitation could be offered in both the statutory and the voluntary sector, but that these should be funded by local authorities rather than central government.

In practice, most of the residential rehabilitation facilities that were established in the late 1960s were run by voluntary organisations, albeit often with statutory financial support. This reliance on the non-statutory sector can partly be related to the long tradition of voluntary work in the rehabilitation of alcoholics. Inebriate hospitals run by Christian groups provided residential treatment for alcoholics from the late nineteenth century onwards.[45] Religious organisations also took up the task of dealing with drug addicts in the 1960s. The Coke Hole Trust, established in 1968, placed emphasis on Christian love and discipline as a way of overcoming addiction; on the ' "family-cum-Christian community", where each of us is made to feel responsible for the other.'[46] Although religion was important to groups such as the Coke Hole Trust, other models were also employed. One organisation, the Rehabilitation of Metropolitan Addicts (ROMA), provided hostel accommodation for drug users and, unusually did not insist that residents be drug free. But, most of the other new centres were orientated towards abstinence, and many of these rehabilitation centres looked to the therapeutic community model to structure their work.

This period saw the emergence of concept-based therapeutic communities in Britain which drew explicitly on American models. Alpha House, established by psychiatrist Ian Christie in Portsmouth in 1968, was the first to open, quickly followed by Suffolk House in Middlesex and Phoenix House in London in 1969.[47] Phoenix House London was in many ways typical of drug therapeutic communities in Britain, as it combined elements of both the concept-based model and the democratic approach. The community was established by Professor Griffith Edwards, head of the Addiction Research Unit at the Institute of Psychiatry, and psychiatrist at the Maudsley Hospital in south-east

London. Edwards had long been influenced by the work of Maxwell Jones, and had set up a therapeutic community for alcoholics-based on the democratic model in the early 1960s.[48] During a trip to the US in 1966, he visited Synanon and Daytop and became friends with Mitchell Rosenthal of Phoenix House New York.[49] Edwards became determined to establish a similar community in Britain and arranged for a psychiatric social worker at the Maudsley, Madeline Malherbe, to spend time at Phoenix House in New York. Malherbe returned to Britain convinced of the value of the Phoenix approach, and also convinced of the need to involve ex-addicts in the establishment and running of the community. She asserted that 'Motivation by ex-addicts and direct treatment by ex-addicts must be retained.'[50] However, finding suitable British ex-addicts was deemed impossible, as none had been through the Phoenix programme or one similar to it, so Edwards and Malherbe recruited two ex-addicts and 'graduates' of Phoenix House New York, Denny and Leida Yuson.[51] Ex-addict involvement was thus built into the project from the outset.

This more active role in treatment for addicts contrasted with the generally constrained position of patients receiving treatment for drug addiction in Britain during the 1960s and 1970s. Following the recommendations of the second Interdepartmental Committee on Heroin Addiction in 1965, controls were placed on those considered to be addicted to heroin or other opiate drugs. All addicts had to be notified to a central authority, as with infectious diseases, and had to attend DDUs if they wanted to be prescribed heroin as part of treatment for their addiction.[52] Drug users were also additionally restricted by the legal system. Buying, selling or possessing a range of psychoactive substances such as cocaine, heroin and cannabis without a prescription had been illegal in Britain since the passing of the Dangerous Drugs Act in 1920.[53] The illegality of drug use further contributed towards the stigmatisation of drug takers, meaning that drug users were even more marginalised and less able to exercise their rights than other mental health patients. A Home Office official told Arnold Trebach, an American Professor of Jurisprudence, that 'Addicts have no rights simply because they are addicts.'[54] Although there was some activism concerned with the legal rights of those arrested for drug offences, as will be seen in Chapter 2, the air of criminality that surrounded drug use acted as an additional barrier to the development of a drug user movement on par with that around mental health in this period. The controls placed on drug users limited their ability to create and run their own organisations; it is significant that all of the concept-based therapeutic communities established in Britain during the 1960s and 1970s to rehabilitate drug users were founded by psychiatrists.[55] However, that is not to say that drug users did not have a role to play in these organisations. Indeed, as an examination of the methods used by Phoenix House in London will demonstrate, this role was important and of lasting significance.

Before Phoenix House in London could open to residents, a suitable property and funding needed to be secured. Edwards and his colleagues at the ARU and the Community Drug Project (another voluntary organisation established by Edwards which will be discussed in detail in Chapter 3) entered into a long drawn out dispute over planning permission for the project, losing one property as a result of local opposition. Eventually, Edwards managed to obtain Featherstone Lodge, a former nurses' home in Forest Hill, south-east London, rent-free from the Maudsley Hospital.[56] Financial support for the project came from a variety of sources. The majority of the cost was borne by the London Boroughs Association, as local authorities were required to support hostel accommodation for vulnerable people.[57] The Home Office also provided funds, supporting a limited number of places in Phoenix House through the probation service.[58] Finally, supplementary benefit for each resident could be claimed and paid to the House to cover their living costs.[59]

Phoenix House London (initially called Featherstone Lodge after the building in which the project was located) opened in December 1969. The project as a whole was co-ordinated by Madeline Malherbe with Denny Yuson as Director and Leida Yuson as Assistant Director. The community was modelled on Phoenix House New York, and like Synanon and the other concept-based therapeutic communities, regarded drug use not as a problem in itself, but rather as a symptom of underlying issues. The project's first annual report stated that, 'An individual's reliance on chemical substances prevents him from reaching solutions to these [underlying] problems. Therefore, while withdrawal from drugs is essential, it is only a preliminary to tackling the problems themselves.'[60] This was to be achieved through the vehicle of self-help: 'The concept of self-help is vitally important in this process. We believe that it is essential for the ex-addict to be given ample opportunity to help himself in his recovery and to assume responsibility for his own life.'[61]

Phoenix House aimed to provide a total therapeutic experience: every aspect of life in the community was designed to be used therapeutically.[62] Time was structured around a rigid daily routine of work (cooking, cleaning, maintenance and gardening) interspersed with educational seminars and group therapy sessions. The work was carried out by hierarchical teams, with new residents at the bottom doing the lowliest job such as washing dishes or peeling potatoes. The hierarchy was designed to allow residents increasing levels of responsibility, but failure to fulfil these responsibilities, or exercise power effectively, resulted in demotion.[63] At all times residents were expected to 'act as if'; 'as if' they lived a rehabilitated life, or 'as if' they were able to tolerate life's daily frustrations without 'acting out'.[64] This was intended to bring about changes in the residents' behaviour, and was enforced through a system of punishments. A minor misdemeanour could result in a 'pull-up' from another resident: 'an indication, simply, that someone has noticed the fault and is concerned enough to point it out and offer a metaphorical hand to help up the one who has slipped.'[65] For more serious 'offences' a resident could be made to wear a sign around their neck indicating their transgression.[66] The ultimate punishment was a 'haircut'. In later years this came to mean a severe verbal admonition, but in the early days of the project could result in an actual haircut, with male residents having their heads shaved, and female residents being forced to wear a stocking-cap. The loss of hair was meant to demonstrate the individual's commitment to the community and to give them something back in return for their concern and support.[67]

Alongside the hierarchy and the use of discipline, the chief therapeutic tool was the Encounter. The Encounter was an unstructured group therapy session led by the ex-addict therapists: 'Anything goes but physical violence in these "no-holds-barred" sessions.'[68] The Encounter allowed residents to voice the tensions of daily life in the community, but was also the place where he or she was encouraged to examine their feelings and address the reasons for his or her drug use. Through mutual aid and self-help the ex-addict would thus be able to rehabilitate him or herself and return to the outside world. 'Confronted by his peers in an atmosphere of gut-level honesty, the former addict learns to see himself as he really is... The human concern of his peers makes change possible.'[69] Self-help and mutual aid were thus at the core of the Phoenix programme.

Life in Phoenix House was unashamedly tough, and only a small proportion of those who showed interest in entering the community actually did so.[70] Referrals came from DDUs, the probation services, GPs, street agencies and even some self-referrals.[71] The only conditions of entry were to be drug-free, and for funding purposes, resident in London.[72] Rehabilitation was supposed to last 18 months, after which the resident would go into 're-entry', a phased re-introduction into society.[73] Successful completion of the programme was termed 'graduation', leaving early was 'splitting'. The word 'graduate' was copied from the American Phoenix programme, which contained a much more formal system of training for those who completed the programme. Using the term 'graduate' clearly implied a degree of professional authority and seems to have been intended to enhance the status of ex-addict therapists. There appears to have been some attempt to 'professionalise' the knowledge of ex-addicts, a move which in some ways contradicted the attitude of the ex-addicts working in Phoenix House London who were, as will be seen, hostile to traditional forms of professional knowledge and expertise.

## 1.3    The relationship between professionals and ex-addicts

Ex-addict involvement in the therapeutic process was at the heart of the Phoenix programme, and the key aspect of the American model applied to British concept-based therapeutic communities. There were three reasons for this. First, ex-addicts were seen as being more persuasive. In a document outlining the need for American ex-addicts in order to establish Phoenix House it was stated that, 'The young person who has been through the whole business of addiction and has left drugs behind may as it seems be a far more persuasive persuader than any social worker with orthodox training.'[74] Second, ex-addicts were regarded as being able to see through other ex-addicts' justifications and excuses. In the words of a visiting psychiatrist connected with Phoenix House during the 1970s, ex-addicts were able to say 'oh, cut the bull shit, I've been there mate.'[75] Finally, ex-addicts were believed to act as a role model, demonstrating that a life without drugs was possible: 'He is living proof of the concept that an addict can be cured.'[76] The value to the Phoenix programme of what is often described as 'experiential knowledge', knowledge gained through experience of living with a condition, was not disputed, but it did sit uneasily with the more professional approach and background of other people involved with Phoenix House.[77] There was considerable tension between the American ex-addicts, the Yusons, and the professionals such as Malherbe, the psychiatric social worker, Mitcheson, the visiting psychiatrist, and others on the project's management committee. In a sense, such discord mirrored long-running tensions between professionals and volunteers within voluntary organisations, but the specific issues within Phoenix House are worth examining in some detail, as they shed light on more fundamental questions about the role of the ex-addict and self-help within Phoenix House and other communities like it.[78]

Conflict between the Yusons and Malherbe beset the project from the very beginning: a dispute even occurred over the name of the community. The Yusons wanted to change the name from 'Featherstone Lodge' to 'Phoenix House London', articulating the community's relationship to Phoenix House in the US much more clearly.[79] Malherbe, on the contrary, wanted to keep the name 'Featherstone Lodge' in order to distance the project from what she saw as some of the excesses of the American model.[80] When the Yusons went ahead and changed the community's name without consulting Malherbe, the true nature of the dispute was revealed. The conflict really revolved around who was in charge of the project, and on what kind of expertise this authority should be built. The Yusons did not want Malherbe to have clinical responsibility for the residents in the community, seeing her more as an administrator. In a letter to Edwards, Malherbe complained that the Yusons 'mistrust and reject my clinical judgement and views and since they are two and I am one, I always get overruled.' She continued 'At times I get the feeling that I am here simply as a smoke screen, a figure head to lend the semblance of professionalism and respectability to a thoroughly unorthodox and risky project.'[81]

For their part, the Yusons stressed the value of their unique experience as ex-addicts and graduates of the American Phoenix House programme. Denny Yuson wrote that, 'The main qualifications for this job, in line with the previous thought, are not credentials, but one's personal definition of who one is and how that can be improved.' The problem, as far as he was concerned, was that 'the project director [Malherbe] is unqualified personally and knowledge wise for this type of work in this concept.'[82] Yet, a psychiatrist connected with the project felt that the Yusons were specifically opposed to professional involvement in Phoenix House, stating that he believed for the Yusons, ' "professional" was a derogatory term.'[83] This strained relationship between the ex-addicts and professionals was partly a function of the American Phoenix House model. Malherbe maintained that 'Phoenix trained graduates are unused to working with professional staff', noting that in Phoenix House New York clinical responsibility was entirely in the hands of the ex-addict staff. For Malherbe, it became necessary to 'assess the relative merits of a programme run by ex-addicts only versus a programme run jointly by ex-addicts with the help of professionals – to me that's the real issue.'[84] However, Malherbe and the Yusons were unable to reconcile their differences, and Malherbe resigned in July 1970.

Malherbe's departure, however, did little to ease the tension between the ex-addicts and professionals involved in Phoenix House. The management committee and the project's sponsors, including the Home Office, became increasingly concerned about some of the methods being used. In May 1971, a Thames television documentary about life in Phoenix House was broadcast.[85] A Home Office official who had seen the programme found a number of aspects of the Phoenix approach, such as the punishments and the Encounter sessions, 'disturbing'. He concluded that, 'While all of this, since there is psychiatric supervision, may be acceptable, it is nevertheless a controversial approach.'[86] But even those carrying out the psychiatric supervision were beginning to have doubts about some of the treatment on offer. A psychiatrist connected with Phoenix House recalled one incident that he had found particularly worrying. One night, Leida Yuson had caught a female resident stealing food from the kitchen. To punish her, Leida called the group together and then forced the resident to stand naked in front of a mirror while Leida told her how revolting she was, and berated her for her behaviour.[87] Another Home Office official noted that it appeared that 'the residents were becoming brain-washed into the acceptance of penance for its own sake.'[88] Some members of the management committee, including Edwards, felt that this was unacceptable, and forced Leida to resign.[89] This resulted in a brief 'mutiny' by the residents, when a group left the house and demanded Leida's re-instatement before they would return.[90] Leida was not re-instated, the residents were persuaded to come back, and though Denny continued to run the community for a few more months, he too was forced out by the end of 1971.[91]

Control of the project was then handed to health professionals. Mike Caldwell, a psychologist, took over clinical responsibilities as Director of Phoenix House. He was followed in 1972 by Dave Warren Holland, a former psychiatric nurse. Under professional control in the second half of the 1970s Phoenix House began to achieve a degree of stability. Some of the harsher treatment methods, such as head shaving, were removed. A more comprehensive system of after-care and support was developed, with the intention of reducing the high rates of relapse.[92] Throughout this period of professional direction, however, a role for the ex-addict was maintained. John Brown, former Navy Captain and Chair of the project's management committee, asserted in a letter to Edwards regarding the staffing of the project that 'it is fundamental to the programme that there shall be some fully trained and responsible ex-addicts on the staff, and this will always be so.'[93] As the community developed, Phoenix House was able to employ some of its own 'graduates'.[94] Faith Miles, a professional working at Phoenix House told Roy Conn from the Inner London Probation and After-care Service in 1972 that 'it is my hope that in the not too distant future Phoenix House will once again have an ex-addict in charge of the community itself.'[95] However, as Brown also told Conn, 'we shall certainly never go back to a position in which an ex-addict is solely responsible for the clinical direction and the professionals have merely "consultative and administrative functions."'[96] Indeed, the project did not get an ex-addict director until 1977, when David Tomlinson took over.[97]

While eventual harmony was achieved between ex-addicts and professionals within Phoenix House, the role of ex-addicts in drug rehabilitation continued to be a contentious issue outside the therapeutic community. In 1977 a row broke out between Phoenix House and Brixton Prison. Phoenix had, for a number of years, sent ex-addicts into prisons to talk to inmates about the programme and to recruit new residents. But, when prison officers recognised an ex-addict Phoenix House staff member visiting the prison as a former inmate, entry into the prison was denied.[98] Following this, ex-addict staff with convictions for drug offences were not allowed into the prison. This policy was strongly opposed by Phoenix House, the project's management committee and the Standing Conference On Drug Abuse (SCODA), the drugs voluntary sector's co-ordinating organisation. Brown complained to the Home Office that 'by far the most successful, indeed I would go so far as to say almost the only successful, persuaders of drug addicts to come to Phoenix House... are former addicts (and so, almost by definition, also ex-offenders).' He argued that, 'The employment of former addicts... is absolutely basic to our work and in very large measure accounts for our success compared with other agencies. If they are not accepted we might as well pack up, at least so far as the Prisons are concerned.'[99] A Home Office official stated that, 'Although sympathetic and promising to make some more general moves towards opening up the prison system to voluntary agencies, I feel that our particular problem lies with the employment of ex-offenders, Brixton's nervousness about drugs in the light of recent problems and the degree of autonomy allowed to Prison Governors.'[100] A meeting was arranged between representatives of Phoenix House, together with SCODA, and Lord Harris, Minister of State at the Home Office.[101] It was finally agreed that a circular be issued to all prison governors making it clear that the Home Office had no 'objection in principle' to the admission of 'ex-offender social workers' to prisons, but Brixton prison still remained reluctant to do so.[102]

This dispute illustrates both the value and the limitations of the ex-addict role in the therapeutic community. Phoenix House was clearly willing to defend the place of the ex-addict in its programme; a further indication of their belief that the ex-addict had a unique and important part to play in the rehabilitation of other drug users. The Phoenix programme placed a high value on the experiential knowledge of ex-addicts, recognising that the ex-addict offered insight into drug use and its treatment not to be found in professional drug workers. This view, far from orthodox at the time, is widely accepted today.

## 1.4   Impact

When attempting to assess the significance of Phoenix House it is important to look at the long-term as well as the short-term impact of the programme. Phoenix House's success in actually keeping addicts off drugs in this period was questionable.

According to Alan Ogbourne and Christopher Melotte, social scientists from the Addiction Research Unit who surveyed the first 100 residents in the project, only a 'minority' of individuals completed the programme.[103] In 1972–73, Phoenix House reported that 69 per cent of residents 'split', and just 3 per cent 'graduated'.[104] Moreover, large numbers of former residents (including those who had successfully completed the programme) used drugs after leaving the house. Ogbourne and Melotte found that only 17 per cent of the first 100 residents were still drug-free in 1974.[105] However, the treatment of addiction was a notoriously difficult area, with all forms of treatment seeing high rates of relapse. Edwards maintained that, 'The fact is that Phoenix has set itself up in business to deal with one of the most difficult problems which a social work project could ever be expected to tackle.'[106] Defining what was a 'successful' outcome was open to interpretation. A progress review produced by the management committee of Phoenix House in 1973–74 noted that, 'The success rate has been low, although this is to a certain extent conditioned by the definition of "success" (e.g. are all splittees failures?).'[107] Ogbourne and Melotte observed that around a third of all admissions, including those who did not complete the programme, remained abstinent or used drugs less intensely than before their stay in Phoenix House.[108]

That Phoenix could have a lasting and possibly beneficial effect on its former residents is further corroborated by anecdotal evidence. A number of drug user activists currently working in the field passed through Phoenix, and have spoken of its impact on their lives. One user activist stated that while her time in Phoenix House was 'terrifying' and the first time she went into the community, 'I just thought this is a nuthouse and split', she went back a few years later and completed the programme. After leaving Phoenix, she and a group of other ex-residents set up one of the first HIV projects orientated towards drug users.[109] Another user activist 'split' from Phoenix House in the mid-1980s, and although he remained critical of some of the methods used, he still considered it to be an option for some drug users, commenting that 'if it [Phoenix House] helps somebody, then that's fine. I know enough people that it has helped.' What the user activist also recognised was the value of the ex-addict staff. He said that although he disliked one of the ex-addict therapists at Phoenix on a personal level, he was a 'Good counsellor though. Really kind of, you know, if you were in a group with him you just sort of thought, well, there's not much point in bull shiting him because he kind of sees.'[110]

According to Ogbourne and Melotte, the majority of those who successfully completed the programme in the 1970s went on to become staff members at Phoenix House, or communities like it.[111] Some of these ex-addict staff also became important figures in the field. Peter Martin, also an ex-resident of Phoenix House, became Director of the charity Addaction. While caution must be exercised about extrapolating too widely from such examples, these cases do indicate that whether or not the programme was a 'success' for a particular individual, it was able to create a place for the ex-addict in their rehabilitation and that of others and also within the wider policy community around drugs. Collectively and individually therapeutic communities like Phoenix House came to have an influence on the development of the field. A group of therapeutic communities (including Phoenix House) played an important role in attempts to co-ordinate voluntary action around drugs through the establishment of the drug voluntary sector's co-ordinating body, the Standing Conference On Drug Abuse (SCODA), in the 1970s.[112] SCODA will be discussed in detail in Chapter 3, but it is important to note here that through SCODA and other organisations, such as the European Federation of Therapeutic Communities, Phoenix House was able to champion the value of the concept-based therapeutic community approach and lobby government for funds.[113] By using these networks Phoenix House was able to gain credibility and access to policy-making circles. A particularly important development was the appointment of David Tomlinson, the first ex-resident director of Phoenix House, to the

Advisory Council on the Misuse of Drugs (ACMD). Tomlinson was the
first open ex-addict to sit on the ACMD, and together with other representatives from the drugs voluntary sector, he was successfully able to
press for the establishment of City Roads, a residential crisis intervention centre for drug users.[114]

The broader impact of Phoenix House (and programmes like it)
can also be seen in its persistence and gradual expansion. Under
Tomlinson's direction Phoenix House took over more buildings at the
Forest Hill site to create a 'campus' and, during the 1980s, also opened
facilities in other locations such as Sheffield, South Shields, the Wirral,
Hove, Bexhill and Glasgow.[115] Phoenix House and the other residential
rehabilitation services underwent a period of financial difficulty in the
1990s, as will be discussed in Chapter 5, but they largely weathered the
storm. Today, Phoenix House (recently re-branded as 'Phoenix Futures')
claims to be the leading provider of care and rehabilitation services for
people with drug and alcohol problems in the UK, and offers community-based programmes and prison-based services in addition to its
residential facilities.[116] Indeed, programmes like Phoenix House are an
increasingly important part of the response to drug use. Since 2006
there has been a concerted effort by the government to provide funding for more places in residential rehabilitation, largely because it has
been shown to be the most successful treatment in terms of getting
users to achieve sustained periods of abstinence from drugs.[117] Other
forms of self-help for drug use would also seem to have grown since the
1970s. Self-help groups based on the 12-step philosophy, such as NA,
have become widespread. In 1979 there was just one weekly NA meeting in Britain, by 1991 there were approximately 223 meetings, and by
2004 there were well over 500.[118]

Where the impact of Phoenix House is perhaps more uncertain,
however, is around the position of the user him or herself. Phoenix
House did create a place for the user in helping themselves and others,
but this was not the starting point for a radical 'user movement' as
was seen with some of the mental health groups founded at a similar
time. While therapeutic communities and self-help as an alternative to
psychiatric treatment had played an important part in the creation of
the mental health users' movement, these concepts were operationalised in a different way in the drug field. Residents and ex-addict staff
in Phoenix House and other concept-based therapeutic communities
remained convinced of their own pathology. Although they believed
that drug addiction was not necessarily a disease in itself, it was
regarded as being indicative of other problems that must be addressed.

This view was re-inforced by the description of the therapeutic process as 'rehabilitation', making the individual 'normal' and returning him or her to society. Such an environment was unlikely to breed radicalism. It is significant that the role created for the user within Phoenix House was for ex-addicts only – this was not about questioning the idea that drug addiction required treatment, or embracing continued drug use as behaviour worthy of defence.

Looking to Phoenix House and communities like it as a potential ancestor of a more radical contemporary drug user movement might, therefore, appear to be to look in the wrong place. Users within therapeutic communities remained largely inwardly focussed, addressing the treatment of individuals, not the wider conditions surrounding drug use. But, by creating a place for the user in their own treatment and that of others, the concept-based therapeutic communities contributed towards the development of another kind of user activism: involvement in treatment. Indeed, it could be argued that the origins of the two different aspects of user activism since the late 1990s – user involvement in treatment services and user activism pushing for radical changes to the drug laws – lie in this earlier period. But, as will be seen in Chapter 2, the challenge to the dominant view on drug use in the 1960s and 1970s came not from health orientated groups such as Phoenix House, but from those attempting to deal with and reform the legal system surrounding drug use.

## 1.5 Conclusion

Phoenix House, and the kind of self-help activity it instigated, was a product of both the wider context in which it was created and the specific individuals and influences involved. The community was rooted in long-standing principles of self-help and mutual aid, but was established at a time when interest in self-help and self-discovery was on the increase. More challenging, activist style modes of voluntarism developed in the 1960s, involving the recipients of health and welfare services – particularly in areas where these individuals had traditionally been denied a role, such as mental health – to a greater degree than in the past. The important role carved out for the ex-addict in the treatment offered at Phoenix House, and the emphasis placed on self-help and mutual aid, was significant not only because it drew on experiential knowledge about addiction but also because it allowed a place for the user in his or her own treatment and that of others. Yet, when the role of the ex-addict in Phoenix House is examined in detail, it is clear that this was tightly managed by the professionals involved with the programme. Opposition to the professional approach to drug addiction by the Yusons was quickly eliminated. Close connections between psychiatrists can be found in the earlier democratic kind of therapeutic community in Britain, and to an extent this tradition was continued in the concept-based therapeutic communities dealing with drug addiction. It is perhaps not surprising then, that the psychiatric view of addiction as pathology persisted within Phoenix House.

This does not mean, however, that Phoenix House should be seen as an essentially professional project which emulated other forms of treatment for addiction found within the statutory sector. Self-help was at the core of the Phoenix programme and it was this emphasis that marked it out from the treatment offered by DDUs. Moreover, by taking on the 'rehabilitation' of drug addicts Phoenix House and other therapeutic communities were attempting to address a different aspect of the drug problem, one that complemented rather than conflicted with the statutory response. In this way, voluntary organisations such as Phoenix House were performing a valuable role: offering a service that the state either could not or would not provide. But, as will be seen in Chapter 2, not all voluntary organisations working in the drugs field had such an easy relationship with the state. Some groups viewed drug addiction not as a problem solely of the individual, but one for which wider society was to some degree responsible. Phoenix House, however, as a result of its grounding in the principles of self-help, saw addiction as something which could only be overcome by the addict him or herself, albeit with the help of their peers. In this way Phoenix House combined the 'old' with the 'new': an old form of voluntarism was thus remade to address a 'new' problem in the 1960s and 1970s.

# 2
# The 'New'? New Social Movements and the Work of Release

The voluntary organisation Release – established in 1967 to provide legal assistance to those arrested for drug offences – was a new organisation seemingly rooted in a 'new' kind of politics.[1] In the late 1960s, according to Geoff Eley, 'The *boundaries* [his italics] of politics – the very category of the political – had been extended by feminists, gay liberationists, environmentalists, autonomists and others.'[2] For what have been described as the 'new social movements' the personal was political: questions of identity and lifestyle were legitimate areas of political protest and activism. This concern with the personal has led numerous commentators to point to the existence of a 'new' kind of politics from the late 1960s onwards. Theorists such as Alain Touraine, Alberto Melucci and Jurgen Habermas posited that new social movements (the civil rights movement, the women's movement, the gay rights movement, the peace movement, the environmental movement and so on) were representative of a 'new' politics: a politics that operated outside Parliament and traditional political institutions, a politics that dealt with new struggles based on identity, culture and lifestyle.[3] For Habermas, this was in contrast to the political conflict which had surrounded capital-labour relations and problems of distribution throughout the earlier decades of the twentieth century. During the late 1960s, he contended, a thematic change occurred in which 'old' politics revolving around questions of economic, social and domestic security was replaced with 'new' politics, concerned with quality of life, equality, individual self-realisation, participation and human rights.[4]

'New' politics resulted in the emergence of 'new' political identities. Social movement analyst Paul Byrne noted that 'people no longer classify themselves just as Conservatives, Socialists and so on, but as "feminists" and "greens".[5] It is often suggested that the creation of these identities was indicative of disengagement with big 'P' politics, that the interest of new social movements in everyday ways of living erased the meaning of the old divide between 'Left' and 'Right'.[6] Feminist scholar Sheila Rowbotham, in the introduction to a book charting the early history of 36 organisations founded during the 1960s, notes that these groups were made up of 'inveterate individualists of the left and the right ... Few refer to any party politics; they were just determined to get things done.'[7] Although social movements of the past had engaged with identity politics, Nick Crossley suggests that what was new about the movements of the 1960s and 1970s was their supposed abandonment of class struggle to concentrate on other conflicts within society.[8] Eley argues that the issues concerning new social movements such as gender, sexuality and personal life were previously marginalised by the Left because these could not be put into a class-political framework. However, he contends 'these were the questions that invaded the Left's imagination after 1968.'[9]

Studies of new social movements and the organisations located within them have frequently pointed to Leftist connections. The association between the Campaign for Nuclear Disarmament (CND) and the Labour Party, for example, has been well documented.[10] That is not to suggest however, that all movements can be typified in this way. Many feminist organisations found the Left unreceptive to their projects.[11] Consumer groups such as the Consumers' Association were strongly committed to party-political neutrality.[12] Yet, the raising of Left-Right connections by commentators on these movements suggests a persistence of 'old' politics within new social movements despite their novelty of protest. As Crossley has observed, within new social movements and their accompanying organisations, the 'old' clearly overlapped with the 'new'.[13]

It is the purpose of this chapter to explore the nature of this overlap between old and new: to examine how one social movement organisation, Release, negotiated their relationship with both 'old' and 'new' politics. Although Release was part of a movement that wanted to create an alternative society to that of the mainstream, its position as a group that offered practical services to drug users in need meant that it was repeatedly drawn into a relationship with the 'establishment'. By concerning itself with questions of identity and lifestyle Release was representative of an interest in 'new' types of political and social problems, but these were often resolved by using 'old' political strategies. Furthermore, the charity faced many of the same difficulties encountered by voluntary organisations in the past, pointing to the continuing relevance of long-running themes throughout the history of voluntarism. Even organisations, such as Release, associated with 'new' types of politics could not, it seems, escape from age-old problems such as the caveats attached to sources of funding.

The continuous inter-play between old and new politics within the experiences of Release between 1967 and 1978 can be observed in four areas: first, the reasons for Release's formation; second, what it did and how effective this was; third, who worked for the organisation; and finally, how it was funded. For Release to be successful they had to interact with 'old' political systems, structures and prejudices. However, the solutions the organisation posed, and the very fact that it attempted to address a different type of problem at all, is more representative of 'new' dynamism around politics and political action. Whether this was indicative of a wholly 'new' type of politics is, of course, more debatable. Craig Calhoun has argued that identity politics can be found within the social movements of the early nineteenth century, suggesting there was nothing 'new' about the 'new social movements' of the 1960s.[14] Yet, when the work of organisations such as Release is considered in detail, it is hard to escape the sense that its members convinced of not only the necessity but also the novelty of their action.

## 2.1  Foundation

In Britain, what has been termed the 'alternative society', the 'underground' or the 'counter-culture' of the 1960s was a loosely defined and sometimes contradictory project. According to Arthur Marwick, this was made up of the 'many and varied activities and values which contrasted with, or were critical of, the conventional values and modes of established society.' There was not, he argues, a 'unified, integrated counter-culture, totally and consistently opposed to mainstream culture' rather, this was made up of 'a large number of very varied subcultures.'[15] The writer and critic Jonathon Green has a somewhat more cohesive view of the alternative society. He asserts that this was an 'educated movement, drawing on the alienated children of the bourgeoisie.' It was not, he asserts

> overtly political, preferring to mock both the right and the left, but it adopted a predictable liberal platform, backing abortion law reform, the abolition of censorship, sexual freedom, banning the Bomb and of course the legalisation of soft drugs.[16]

The alternative society was, therefore, concerned with the kinds of issues surrounding identity and lifestyle that have been described as being features of the new social movements. Counter-cultural protest took the form of the adoption of different styles of dress, hair, music, film, literature and art, but there was also an attempt by the underground to develop alternative institutions to those of the mainstream, dominant culture. These included the Free School (later the Anti-University), Arts Lab, the Indica bookshop, *International Times* and Release, described by one of its founders, Caroline Coon, as 'the welfare branch of the alternative society.'[17] Release, according to Marwick, was 'one of the great sixties successes', and for Green 'one of the most significant legacies of the alternative society.'[18] It is certainly the only one that has lasted: Release still exists today, whereas many other institutions of the alternative society disappeared by the mid-1970s.[19]

The use of illegal drugs, and cannabis in particular, was a celebrated part of the counter-culture.[20] The numbers of those taking cannabis increased dramatically during the late 1960s as use of the drug moved beyond a small group of mainly West Indian users to the much larger underground scene: there were just 626 convictions for the possession of cannabis in 1965, compared to 2,393 in 1967.[21] According to the infamous pro-cannabis advertisement that appeared in *The Times* in 1967 the drug was taken 'for the purpose of enhancing sensory experience.'[22] Hippie culture, observed sociologist Jock Young, was essentially 'Dionysian', directed towards 'sexual pleasure, physical euphoria and enjoyment.' This emphasis on hedonism, he contended, brought the underground into conflict with mainstream society as it undermined the work ethic upon which Western industrial societies were based.[23] The signatories of *The Times* advertisement may have argued that 'the law against marijuana is immoral in principle and unworkable in practice' but any defence of cannabis use was more than matched by tabloid newspaper condemnation of the drug and all those who took it.[24]

Release was established in 1967 partly as a response to these attacks on cannabis use. The founders of Release, two art students in their early twenties, Caroline Coon and Rufus Harris, met after a demonstration against the publication by the *News of the World* of a virulent assault on Mick Jagger and his use of drugs. Harris told Coon that there was to be a meeting of members of the underground to discuss what could be done about the injustice of the drug laws and the escalating number of arrests for drug offences.[25] Coon had particular reason to believe that the system was unfair: her Jamaican boyfriend had recently been sentenced to two years imprisonment for the possession of a single 'joint' of cannabis.[26] This gave Coon some experience of the legal world and also the inspiration for the name of the organisation – 'Release' – because she was trying to get her boyfriend released from prison.[27] Although other counter-cultural figures such as the rock entrepreneur Joe Boyd and artist Michael English were involved at the outset, Release rapidly became Coon and Harris's project.[28] Initially run from Coon's basement flat, she and Harris stated that Release was 'established to help those who have been arrested for drug offences.'[29] Release provided a 24-hour emergency telephone service and produced thousands of 'Bust Cards', informing people of their legal rights.[30]

For Coon, it was this framework of rights that informed the foundation of Release. She stated that: 'For me, Release was not about drugs *per se*... For me Release was essentially about civil liberties, legal rights, and what we now call human rights.'[31] The significance of civil, human, legal, patient and consumer 'rights' increased in this period. Partly inspired by the civil rights movement in America other movements and organisations began to apply the language and rhetoric of individual and collective entitlement to a range of scenarios.[32] For groups such as Amnesty International, established in 1961 to campaign for the rights of prisoners of conscience, the notion of human rights as outlined in the United Nations Declaration of Human Rights (1948) was clearly crucial, but a broader interpretation of 'rights' began to take shape as consumer 'demands' became increasingly potent.[33] A slippage between the rights of the 'citizen' and the demands of the 'consumer' can be observed in organisations such as the Patients' Association (1963) the Consumers' Association (1956) and the re-orientation of groups such as the National Association of Mental Health (later MIND) away from mental health professionals and towards mental health service users.[34] The intricacies of this relationship is not the focus of this chapter and requires further investigation, but the notion of 'rights' for dealing with the problems of 'quality of life, equality, individual self-realisation, participation and human rights' representative, for Habermas, of 'new' politics, appears significant.[35] The notion of rights, for example, was important for the environmental movement (the right to clean air and water) and for the women's movement (the right to equal opportunities).

By campaigning for the legal rights of the drug user, Release was thus part of a much broader movement surrounding individual and collective entitlement. Release was not just about drug taking; indeed, Coon asserts that when the organisation started she did not drink or smoke and had never taken drugs.[36] That is not to say, however, that drugs are not important in developing an understanding of the work of Release. Examining their activities more closely would indicate that Release became involved in many other key issues of the 1960s and 1970s, those rising from 'new' forms of politics concerned with identity and lifestyle. However, in order to be successful in delivering their message and providing assistance to those perceived to be in need, Release utilised 'old' political methods, exploiting mainstream as well as counter-cultural connections.

## 2.2   The work

Staffed initially by volunteers, including Coon and Harris themselves, Release was open from 10am until 6pm, and on two evenings a week legal and medical advice was available from a team of solicitors and doctors, who also worked on a volunteer basis.[37] Release expanded rapidly and by the end of 1967 it moved from Coon's flat to offices on Princedale Road, in Notting Hill.[38] No figures exist for the number of people who contacted Release in 1967, but the organisation recorded seeing 603 cases in 1968 and 661 cases in 1969. There was a considerable jump in the number of cases seen by 1970, when 3,846 were recorded, which amounted to around 74 cases a week.[39] A large proportion of these related to arrest for drug offences, usually the possession of cannabis, but Release also saw a considerable number of young people with other problems, including unwanted pregnancies, medical and psychiatric conditions, homelessness, difficulties with housing and other legal concerns.[40] Jeremy d'Agapeyeff, a sociologist brought in to asses the work of Release in 1972 noted that

> Release's original aim was to help with the drug problem; later, of necessity, through following the demands of the community it served, its sphere of activity expanded to include the whole spectrum of problems encountered by the so-called alternative society.[41]

For those working at Release, part of their function as the 'welfare branch of the alternative society' was to act as a 'buffer' between the mainstream and the counter-culture.[42] This involved making legal and medical services available to a group of people whose lifestyles and values made it difficult for them to use 'straight' services.[43] D'Agapeyeff noted that 'Release not only functions to plug gaps in the welfare service but it also acts as a bridge between subcultural groups and the establishment.'[44]

Supplementing existing statutory welfare services had long been seen as one of the roles of the voluntary sector. But, as was noted in the Introduction, this function seemed to acquire greater significance during the 1960s and 1970s as needs, and the perception of these, began to change. The voluntary sector as a whole expanded considerably. Between 1961 and 1971 the number of registered charities rose from 1,182 to 76,648. This dramatic increase can largely be accounted for by the introduction of more efficient methods for registering charities, but it is estimated that around 10,500 of these were entirely new organisations.[45] Some of these organisations such as Child Poverty Action Group and Shelter were concerned with the 'rediscovery' of poverty in the 1960s.[46] Others, including Release, the Campaign for Homosexual Equality (1969) and the Gay Switchboard (1974) were connected to wider social movements, campaigning on the behalf of what Beveridge called the 'distressed minorities'.[47] For both types of organisation there was a related awareness that the statutory sector was failing to provide adequate services in a number of areas.[48] Voluntary groups emerged to fill in the gaps. Harris has stated that Release was doing 'pioneering social and community work', but the services Release offered were not necessarily that different to those found within the statutory sector.[49] Where the organisation differed, according to social psychologist and Release trustee Michael Schofield was that the advice it gave came from a more trusted source. Release was an 'organisation run for the young by the young.'[50]

Release helped many young people with a large range of problems but the area in which they could claim the most success was in legal advice, particularly in cannabis cases. In an analysis of cases seen in 1970, Release noted that individuals who sought their advice after arrest were less likely to be charged with an offence than those who had not: the Wootton report on cannabis stated that 25 per cent of arrests for possession of the drug led to charge, compared to 9 per cent of the cases seen by Release.[51] If a client was charged, Release's lawyers were often able to keep him or her out of prison. Wootton found that 517 out of 2,734 (19 per cent) of convictions for cannabis offences in 1967 led to imprisonment, whereas just 6 out of 603 (less than 1 per cent) of all legal cases (including non-drug offences) seen by Release in 1968 resulted in a prison sentence.[52] Schofield noted that, 'One of the best achievements of Release has been to build up a panel of lawyers who [were] skilled in the problems of drug cases.' This was significant, he argued, because if 'a policeman arrests you, even if you are absolutely innocent, the only way you can be sure of staying out of prison is to employ a solicitor.'[53]

Yet, Release did more than keep people out of prison: defending the legal rights of the drug user was part of their desire to highlight the fact that the ' "drug problem" ... cannot be solved by the conventional means of criminal reprimand.' They felt that 'medical or social solutions [were] more likely to be successful.'[54]

For this reason Release continued to be involved in campaigning, not just for the reform of drug laws but on other matters too. In addition to agitating for the legalisation of cannabis, Release actively supported the woman's right to choose, protested against the Vietnam War and campaigned for squatters' rights during the 1970s.[55] But, it was on drug issues that the organisation was particularly prominent. Release lobbied for improvements in the treatment facilities provided for heroin addicts, and along with other street agencies, highlighted the growing problems posed by barbiturate and poly-drug use which was largely ignored by NHS Drug Dependence Units.[56] This message was communicated through numerous publications produced by Release, including a newsletter, and in 1969 *The Release Report On Drug Offenders and the Law*, which detailed the organisation's work and attacked police corruption in relation to drug arrests.[57] Release was also able to lobby government directly when asked to give evidence to the Advisory Council on Drug Dependence. They submitted written evidence to both the Wootton Committee on amphetamines and LSD and the Deedes Committee on police powers of search and arrest in 1969.[58] In addition, Release was in direct communication with influential civil servants in the drugs field, such as the Chief Inspector of the Home Office Drugs Branch, Bing Spear. Coon found Spear 'particularly helpful' and she believed that the Home Office 'in principal agree with what we are trying to do.'[59] This may not necessarily have been the case; Spear stated that the Drugs Branch had a 'long-established policy of keeping in contact with anyone working in the drug dependence field' which did not 'imply approval, or disapproval' of the organisation concerned.[60]

Indeed, being heard was not the same as being listened to. Public and political attitudes to the issues on which Release campaigned did change, but very slowly. It is almost impossible to quantify how influential Release was in achieving these changes. As Bruce Wood has observed in relation to the work of patients' associations: 'Measuring effectiveness in any context is always tricky', and it could be suggested that in a field such as drug policy, which was strongly politicised and deeply divided, even more so.[61] The negative reaction to the publication of the Wootton report on cannabis in 1968, which suggested that possession of small amounts of the drug should not result in imprisonment, highlights the extent to which recommendations believed to be unacceptable by politicians and the media could be disregarded, even when made by a group of 'establishment' experts.[62] Furthermore, until the 1980s, the formulation of drug policy was dominated by the medical profession, largely to the exclusion of other agencies.[63]

In this climate it was difficult for Release to have a major impact on policy-making, but that does not mean they were insignificant or ineffective. The organisation, Green noted, gained the respect of others working in the field, including, rather grudgingly, the police.[64] Coon remarked that 'despite our youth' Release was regarded as 'a credible, serious organisation.'[65] It would seem that she had a vision from the outset that in order to effect change Release must make contact with the 'establishment.' Coon stated that 'I knew that to start something like Release I had to go to the top', so she made an appointment with the head of Scotland Yard. Coon recounts how she was handed over to the director of public relations:

> There he was behind his desk scowling. It was for me a classic moment. I realised that the establishment had assumed that Rufus and I were the stereotype of what they considered "disgusting" hippies: smelly, barefooted, working class ne're-do-wells! Well, of course, I walked up to Mr PR's desk, put out my hand, introduced myself and said, "How do you do." Mr PR leapt to his feet and said "Oh, oh, oh, good gracious me, ah, Miss Coon! Would you like a glass of sherry?" It was very funny.

Coon attributed the policeman's reaction to her upper-class accent and demeanour, this gave her, she asserted, 'a poise and manner which got my foot in the door with the leaders of the establishment.'[66]

## 2.3 The workforce

Coon was clearly quite prepared to use her upper-class background to publicise Release's message, and this helped her in her role as their spokesperson until she left the organisation in 1971. However, her background and connections were less beneficial for her relations with parts of the alternative society and even some of those who worked at Release. Steve Abrams, a key figure in the legalisation of cannabis campaign and prime mover behind the 1967 *Times* advertisement, said that 'Caroline did a magnificent job for Release' but

> There was a tremendous hostility towards her, and the better she was at it, the more they [the underground] didn't like her...the fact that she had a rather posh accent and the fact that she was a woman led to disaffection of large segments of the male underground towards her.[67]

Despite being 'alternative' the underground was deeply misogynistic.[68] Self-confessed 'hippie chick' Nicola Lane felt that the sixties were 'totally male dominated.' She said that, 'A lot of girls just rolled joints – it was what you did while you sat in quietly in the corner, nodding your head. You were not really encouraged to be a thinker. You were really there for fucks and domesticity.'[69] Despite being 'alternative', the counter-culture was clearly imbued with the 'old' politics of gender inequality.

It is hardly surprising then, that there should have been a degree of hostility towards Coon and this appears to have been present even inside Release itself. An oblique critique of Coon, her gender and her supposed 'elitism' appeared in an article in Release's Newsletter, *Connection*, in 1972. The author stated that, 'The days of charismatic Caroline and her ladies are gone. Yet to a large extent their legacy remains – for good or for ill.'[70] Five years later, in a publication produced to mark Release's tenth anniversary that sentiment seemed to have persisted. There was, Release worker Roger Lewis argued, some truth in the assertion that in the early days 'Release people were a group of Lady Bountifuls dispensing help and support like a nineteenth century pietist [sic.] charity.'[71] How ever ill justified, the allusion to 'Lady Bountiful' would suggest that questions were being raised within Release about the nature of the organisation with reference to much older debates about philanthropy and charitable work. 'Lady Bountiful' was a largely pejorative term used to describe upper and middle class female philanthropists during the nineteenth century. Women were key figures in many charitable organisations in this period, but their work was often, and has since, been criticised as being patronising towards the working-classes.[72] Philanthropy, it was argued, denied the working-classes agency and prevented them from forming their own organisations based on principles of self-help and mutual aid.[73] Charity was thus a form of social control, imposing bourgeois values onto the working-classes in order to preserve order and capitalism. However, more recent scholarship on voluntarism in the nineteenth century has suggested this can be seen as the basis of consensus, bring together the middle and working classes in the creation of a peaceable kingdom based on values of decency, independence and disgust of the so-called 'undeserving' poor.[74] In this picture, philanthropy and mutual aid are drawn much closer together, united in a common goal.

In the case of Release, 100 years or so later, the distinction between philanthropy and mutual aid appears even less pronounced. Release was frequently accused of becoming 'elitist', distant from the rest of the alternative society and removed from the people it was supposed to help.[75] Yet, Release continued to regard itself as a self-help group. At its tenth anniversary in 1977 Release stated that it was an 'alternative organisation within which the workers tend to be indistinguishable from the people they are assisting.'[76] In a sense, this was wholly accurate. It is likely that Release was a predominately middle class organisation helping a disproportionately middle class clientele. An analysis of cases seen in the spring of 1970 found that 59 per cent of the individuals seen were middle class and 41 per cent what was termed 'lower class'.[77] This was disproportionate when compared to national surveys which repeatedly found that around 67 per cent of people in Britain considered themselves to be working class and 29 per cent middle class.[78] A later study of Release's cases conducted in 1972, using the Registrar General's Classification of social class and taking occupation, education and father's occupation as determining factors, found that 52 per cent of individuals seen were middle class, and 48 per cent working class. Despite this more even demographic, the study's author felt it necessary to state that this 'should help to discourage the idea of Release as a self-indulgent middle-class organisation', suggesting this was often how the charity was perceived.[79]

It has been suggested that the alternative society more generally was a largely middle class movement, and the middle class bias of Release was typical of many organisations founded in this period.[80] For instance, Des Wilson, the first director of homelessness charity Shelter, asserted that the organisation's founders were 'representative of the middle class paternalism that had characterised charity since Victorian days.'[81] Finlayson noted that the majority of voluntary social action during the 1960s and 1970s was carried out by the middle classes, 'there was a lack of any real working-class involvement.'[82] Studies of new social movements have also repeatedly highlighted an over-representation of the middle classes.[83] Frank Parkin's classic analysis of CND in 1965–66 found that 83 per cent of the adult sample and 62 per cent of the youth sample were professional, managerial or 'white collar' workers.[84] More recent studies of CND display a similar pattern, one that was also replicated in numerous environmental groups.[85] This has led some commentators to assert that new social movements were largely representative of middle class interests.[86] Others, such as Paul Bagguely, argue that assessing the social base of a movement is not necessarily the best way of measuring the interests it represents, but it is hard to escape from the conclusion of Claus Offe that new social movements were exemplar of the 'politics of class' even if they were 'not on behalf of a class.'[87] This is not something explicitly dealt with by many early theories of new social movements, including that of Habermas, and for Crossley represents a serious omission 'because it suggests that the "old politics" of class remains alive within the "new politics".'[88]

Even after Coon, and later Harris, left Release the organisation seemed to retain a strongly middle class, and increasingly professional, orientation. In 1972 four of the eleven full time staff were graduates, including key figures such as Desmond Banks, Release's librarian and Don Aitken, the information officer.[89] Some of those working at Release had relevant professional or vocational qualifications and experience. By 1977, in response to increasing demand for legal and medical services, the organisation employed three qualified solicitors and, somewhat sporadically, a psychiatric social worker.[90] Such developments can be seen as being representative of two trends: the growth of social work and the social worker in the statutory sector, and also the increasing professionalisation of many voluntary organisations. Social work had been growing as a discipline since the late 1950s and expanded still further in the 1960s as spending on personal social services doubled between 1960 and 1968.[91] Social work also began to assume greater visibility and respectability. In 1965 the Seebohm Committee recommended that each local authority establish a social service department to unite social workers and co-ordinate service provision.[92] This increasing professionalisation of social work within the statutory sector had an effect on similar kinds of work being undertaken in the voluntary sector. As Tanya Evans has shown with respect to voluntary organisations working around the issue of poverty, the expansion of social work as a profession and the parallel growth of social administration and sociology as academic subjects was important not only for the establishment of voluntary groups in this area, but also for their overall direction. Poverty groups increasingly employed graduates and those with professional qualifications so that over the course of the 1970s these 'became professional rather than voluntary organisations.'[93]

A somewhat different picture can be found with Release. Although the organisation did employ some graduates, most staff during the early 1970s had no formal training in either social work or counselling. This was important for Release, according to d'Agapeyeff, as the charity would lose credibility with its client base if the workers adopted the skills and techniques of trained social workers. Yet, by acting as a 'buffer' between the client and the 'establishment' Release staff were repeatedly brought into contact with the police, the courts, the probation service and other key elements of mainstream society. As Coon commented, 'We were absolutely alongside establishment culture.'[94] What is also significant about the staff of Release was that paid employees worked together with unpaid volunteers from as early as 1969.[95] In 1970 there were nine paid staff working at Release and ten volunteers, by 1971 there were 15 staff and over 20 volunteers.[96] Although Coon stated that employees were paid the 'bare minimum' (between £15 and £18 a week during the early 1970s), staff wages were still the organisation's greatest single expense.[97]

For the voluntary sector as a whole the payment of salaries was, and to some extent remains, a major issue. Voluntary work is often assumed to be unpaid. But, as Jos Sheard has pointed out volunteering in this sense is not necessarily synonymous with the voluntary sector. Voluntary organisations may rely on the efforts of unpaid volunteers but many also employ paid staff.[98] In response to this David Billis has constructed a model of the voluntary sector which places organisations into the categories of either 'unambiguous' or 'ambiguous.' Unambiguous organisations are supposedly 'pure', comprised solely of unpaid volunteers, whereas ambiguous organisations may employ paid staff.[99] These are not just matters that have concerned academics, but were also important for voluntary agencies on the ground. For Release the issue was not so much the payment of salaries and the potential this had to undermine their 'pure' 'unambiguous' voluntarism, but the dilemmas associated with greater professionalisation. The payment of staff placed Release within what Billis describes as the 'bureaucratic world', where distinctions are made between employer and employee, between paid worker and volunteer, and between counsellor and client.[100] Within Release no such distinction was supposed to exist; the organisation wanted to be no different to those it helped, but the payment of staff inevitably marked a divide between worker and client.

For Release's critics this was further evidence of the 'corruption' of the organisation. D'Agapeyeff reported that some members of the underground felt by 1972 that Release had become 'corrupted, divorced from the community it is supposed to serve and part of the establishment.'[101] However, at the same time Release, like many other voluntary organisations, did not always conform to the bureaucratic standards required by those in authority. Release, d'Agapeyeff argued, found itself in 'an isolated position; too "straight" for sections of the underground, yet too "chaotic" or unstructured for the more traditional social welfare services.'[102] This encapsulates a classic dilemma experienced by many voluntary organisations about whether to remain true to their radical origins or to move closer to the state in order to gain greater authority and stability. This was a problem most clearly been replicated in issues surrounding funding.

## 2.4  Funding

Release was always dependent upon a mixture of sources for its funding, but the evidence points to a gradual transition from counter-cultural sources of revenue to more established sources of funding, and ultimately government grants. Initially funds were raised by the donation of 6 pence from the sale of each admission ticket to alternative clubs such as UFO and the Electric Garden. What Green described as 'tithes' were, according to Coon, 'like the alternative society's community tax.'[103] Large donations of several thousand pounds each were made by key counter-cultural figures such as George Harrison and Eric Clapton.[104] Numerous smaller donations came from private individuals, student unions and companies such as Virgin Records, the Whitbread brewery and Marks and Spencer.[105] Release even encouraged people to send in Green Shield Stamps and cigarette coupons.[106] More glamorous sources of revenue came from benefit concerts such as that given by Elton John in December 1973 and other events, including the premiere of Mick Jagger and James Fox's 1970 film *Peformance*.[107] Gifts were also donated to Release such as prints by artist David Hockney and some more bizarre objects including a gold-plated sculpture of Polish film director Roman Polanski's penis.[108] Despite these diverse sources of revenue Release was almost perpetually in financial crisis. Coon stated that Release went bankrupt in 1971, forcing her to abandon her studies and raise funds for the organisation full time.[109] For the years where reliable figures exist, from 1972 to 1975, Release's accounts show the organisation made an annual loss of a few hundred pounds each year.[110]

Counter-cultural sources of revenue were clearly insufficient. Disgraced former War Minister John Profumo, who did a considerable amount of charity work once he left the government, advised Release that it was 'very important to become a Registered Charity as soon as possible. I know this will help, not only with funds, but in every way.'[111] Release acquired charitable status in 1971 after a lengthy application process.[112] This did appear to make fundraising from more orthodox sources easier. Release received a grant of £1,000 from the Rowntree Trust that same year, and later a much larger grant which enabled them to employ d'Agapeyeff to assess their work in 1972.[113] Accepting money from these kinds of sources was a contentious issue for those working at Release. An article in their newsletter in 1972 reflected:

> We have always been wary of taking money from the state yet we accept it gratefully from Trust funds established by business phi-lanthropists and from individual donors. We emphasise that there should be no strings attached to donations and ostensibly there are none. However, the strings of the good puppeteer can be as restrict-ing as the manacles of a jailer.[114]

Indeed, becoming a charity did place additional restraints on the campaigning activities undertaken by Release. Charity law set down that any activity judged to be 'political', attempting to change political structures or policies by pressure or propaganda, should not be considered 'charit-able'. Charitable status could only be granted if the primary purpose of the organisation was 'non-political', although groups were granted char-itable status if their campaigning activities were ancillary and subordin-ate to other activities.[115] Some groups such as Amnesty International created a separate 'Prisoners of Conscience Fund' which was a charity while the main organisation remained non-charitable.[116] However, other organisations such as Child Poverty Action Group and Age Concern were able to continue lobbying government once they became registered charities as this remained 'subordinate' to other work.[117]

For Release, the potential limits to campaigning after achieving char-itable status appears to have been less significant than the possible con-sequences of accepting money from the state. Release had been invited to apply for a grant from the Home Office in the past, but according to their newsletter 'declined after much debate'. But, in May 1974, 'as we plunge into the horrors of yet another financial crisis, the matter is rearing its tempestuous head again.'[118] In June, the newly established Voluntary Services Unit (VSU), part of the Home Office, offered Release a grant of £20,000 a year for three years.[119] Coon asserted that finances were in such a bad state that the organisation could not survive without government funding.[120] It was also noted that the charity's trustees felt

Release would 'have less difficulty in getting grants from foundations if [they] were receiving government money.'[121] Release accepted the grant in November 1974, but not without reservations.[122] Coon, who no longer played a day-to-day role in the running of the organisation but remained involved as a Release trustee, wrote to the charity warning them that 'money from such an established source always tends to make [an] organisation lose contact with the community and grass root vibes and struggles.'[123] She was glad that Release was to receive the money, but recommended that they continue to raise at least 35 per cent of their income from other sources. Coon hoped that 'every endeavour will be made by Release to remain and develop as a biting, controversial pressure group.'[124]

The effect that government funding had on groups such as Release with a campaigning function, and the increasing role played by the state in financing the voluntary sector more generally, has been a matter for extensive debate. Statutory funding of voluntary groups was suggested as far back as the early Edwardian period, but a more concerted effort on the part of the state to resource voluntary action can be discerned since the 1970s.[125] The creation of the VSU in 1973 and the publication of the Wolfenden report on *The Future of Voluntary Organisations* in 1978 represented a greater interest by the state in the potential for the voluntary sector to take on a greater role in welfare provision.[126] This was desirable for two reasons. First, because greater utilisation of the voluntary sector was regarded as being more cost-effective in a time of fiscal crisis, and second, particularly after Margaret Thatcher's election victory in 1979, because this could help 'roll back the state': greater use of the voluntary sector would diminish the need for statutory services.[127] Release's acceptance of government money thus placed it squarely within a wider drift towards statutory funding of voluntary action, but this did raise a number of issues, particularly about the continued independence of voluntary groups. More recent government funding of the voluntary sector has concentrated on inviting voluntary organisations to tender for contracts to provide a particular service. These come with all sorts of conditions attached which, as Jane Lewis has observed, are often perceived by voluntary organisations as threatening their autonomy and limiting their ability to carry out campaigning work. Completely unconditional grants, she noted, were extremely rare by the mid-1990s.[128]

However, this does not appear to have been the case in the 1970s. Release was not subjected to any additional requirements after accepting Home Office money in 1974. Other voluntary organisations working in the drugs field, such as the Institute for the Study of Drug Dependence (ISDD), also received large government grants without any conditions attached. Like Release, the ISDD did, however make sure they continued to bring in other sources of revenue. This was important, recalled former ISDD Director Jasper Woodcock, because it enabled them to preserve a degree of independence and resist government interference.[129] Looking at the work of Release after 1974 would suggest that they too were able to maintain their autonomy despite statutory funding. Release continued its sometimes controversial work, maintaining a stance that differed widely from government policy on a range of matters. They were critical of NHS Drug Dependence Units and the treatment offered for heroin addiction.[130] Release also persisted with their support for the legalisation of cannabis, offering financial assistance to a re-invigorated campaign that began in 1978.[131] The organisation continued to cause controversy with their publications such as the *Release Guide to Hallucinogenic Mushrooms* which offered advice on the safer use of drugs.

Although such a harm minimisation approach to drug use was later to become government policy in the wake of HIV and AIDS during the late 1980s, in the previous decade this was felt to be tantamount to encouraging drug taking. This was certainly the tenor of the response to Release's opening of a drug telephone hotline in 1975 designed to provide 'objective, non hysterical information' to drug users.[132] Conservative M.P. Alan Clarke believed the recorded information line was likely to encourage young people to take drugs rather than avoid them.[133] *The Times* reported that the service had been attacked by Sergeant Leslie Male, Chairman of the Police Federation, for giving out information on the prices users should expect to pay for their drugs. Release was not, the policeman contended, trying to 'stamp out' drug taking, a position he regarded as inappropriate for an organisation receiving public money.[134] Accepting government funding did, therefore, open up Release to attack, but it did not prevent it from attempting controversial work. The service attracted so much adverse publicity that it was withdrawn in the summer of 1975.[135] Repeated attempts to have it re-instated failed; the Post Office claimed it was prohibited by law from airing controversial messages, but the implication was that the Home Office and the Department of Health had influenced the decision.[136] It is interesting to note that if this was what happened, the Home Office exerted pressure on the Post Office and not Release to abandon the hotline, a move that would suggest Release was able to preserve its independence despite statutory funding. So, while taking government money had the potential to

undermine Release's autonomy and their radical stance this did not necessarily appear to be the case. Indeed, though receiving a Home Office
grant in 1974 was obviously a significant moment for Release it was not
necessarily at odds with the organisation's general relationship to the
state before this date. A much closer association existed between Release
and the 'establishment' than might be expected for an organisation that
sprang up from the 'alternative' society.

## 2.5   Conclusion

The surprisingly close relationship between mainstream and alternative
exposed by an analysis of the funding of Release is a pattern repeated
in other areas of the organisation's experiences between 1967 and 1978,
as this chapter has demonstrated. It was clearly an amalgam of old
and new politics that influenced the activities of Release and shaped
the framework in which the organisation operated. To an extent, this
position was unavoidable; a direct consequence of Release's desire to
act as a 'buffer' between the alternative society and the establishment.
This role brought Release into regular contact with mainstream authorities and in order to operate successfully in such a world, Release had to
engage with it. Crossley, in his examination of the work of the National
Association of Mental Health (NAMH), developed Bourdieu's concept
of a 'field' to describe a similar process. He argues that NAMH were
able to get their message heard because they had the appropriate skills
and social capital for effective action in a number of important 'fields';
most notably the mental health 'field' and the parliamentary 'field.'[137]
Release too, had, or was able to develop, the relevant skills to interact
with a range of other 'fields' outside the alternative movement in which
they were located.

   This did not, however, detract from the novelty of their work. Indeed,
by attempting to provide an alternative to mainstream services Release
were taking the alternative society's project, and that of new social
movements more generally, to its logical conclusion. If Phoenix House
took an old form of voluntary action and remade it to deal with a new
problem in the 1960s and 1970s, Release took a new approach to this
new problem, but were forced to resort to old ways of getting their message heard. To explain this, it is necessary to look to the relationship
between voluntary organisations and the state, a theme explored in
Chapter 3.

# 3
# Drug Voluntary Organisations and the State in the 1960s and 1970s

As the previous chapter demonstrated, even organisations like Release, which attempted to provide a new approach to what they saw as a new kind of problem, could not escape the 'old' politics and institutions which surrounded them. The character of the relationship between different drug voluntary organisations and the world in which they were located is an important topic, and requires further consideration. Whether 'old' or 'new' in their outlook, voluntary organisations working in the drugs field did not exist in a vacuum, but were instead connected to a number of other bodies, groups and communities. Voluntary organisations at the local and the national level interacted with communities, with drug users themselves, with medical professionals, with drug researchers and with the state. Examining the nature of this relationship helps to throw light on some of the more general issues that surrounded voluntary action during the 1960s and early 1970s. Looking at how various voluntary organisations regarded the existing statutory services for drug users, and in turn, how these voluntary groups were seen by these statutory services, reveals much about the nature and wider purpose of voluntary action in this period. The significant upswing in voluntarism from the 1960s onwards has often been explained in terms of the failure of the comprehensive welfare state. There was, according to Finlayson, a 'certain waning of the initial optimism that the state alone would solve all of the problems of society.'[1] Statutory welfare services were thought to be becoming remote, bureaucratic and unable to meet the needs of its citizens sufficiently. Voluntary organisations emerged to 'plug the gaps', providing services where the state did not.

This general picture is replicated in the drugs field. Organisations were established to cater to different aspects of the drug problem not being addressed by the existing statutory services. Local community-based groups, known collectively as 'street agencies', formed to provide services for drug users including counselling, advice and sometimes simply a 'place to be'. But, as a more detailed examination of the interaction between these groups and the statutory drug services (in the form of the Drug Dependence Units, DDUs) will reveal, this was something of an ambivalent relationship. The role performed by the street agencies often complemented that of the DDUs, with voluntary groups such as the Community Drug Project offering services that statutory organisations were unable or unwilling to provide, but there was also a degree of conflict between the DDUs and the street agencies. Some DDU doctors thought that the street agencies actually encouraged drug use by perpetuating 'a drug subculture'.[2] Street agency workers were not any more approving of DDUs, arguing that these rarely provided meaningful treatment.[3] Underpinning this argument was a deeper disagreement about the nature of the drug problem. Many voluntary organisations working with drug users at street level believed that drug use was essentially a social problem. According to the drug voluntary organisations' co-ordinating body, the Standing Conference on Drug Abuse (SCODA), 'Whilst there may be physical complications arising from drug use which require a medical response, there is no logical reason to suppose that drug use of itself is a medical problem.'[4] In contrast, doctors working at the DDUs tended to view drug addiction as a medical problem with social implications rather than the other way round.[5]

Despite the tension between the DDUs and the street agencies, almost all of these voluntary groups received some funding from statutory sources. Various branches of both local and national government supported local, community-based groups and also national organisations such as SCODA. This suggests that the state recognised the value of these groups, perhaps not in spite of, but because of, their different approach to drug use. Throughout the 1960s and 1970s there was growing statutory interest in the work of voluntary organisations. Reports assessing the role of the voluntary sector published in this period, including the Aves report of 1969 and the Wolfenden report of 1978, pointed to the importance of voluntary organisations and their ability to complement, supplement, extend and influence statutory provision.[6] Financial support for voluntary organisations from statutory sources increased in the drugs field and increased more widely. Yet, stronger financial ties to the state also raised the possibility of diminishing independence, something many voluntary groups resisted. SCODA, for example, wanted its relationship with the state to be based on 'partnership not patronage.'[7]

To a great extent, this desire appears to have been realised. Voluntary organisations in receipt of statutory funds were afforded a considerable degree of independence throughout the late 1960s and early 1970s. Moreover, taking money from the state actually seems to have enhanced the reputation of some agencies and allowed them access into policy-making circles. Representatives from voluntary organisations sat, for example, on the Advisory Committee on the Misuse of Drugs (ACMD), the key governmental advisory group on drug policy. Indeed, by the end of the 1970s there were signs that statutory and voluntary agencies were increasingly working together in addressing the drug problem. Projects such as the City Roads Crisis Intervention Centre combined statutory and voluntary approaches to provide a 'hybrid' service.

All of this would suggest that the support provided by the state to voluntary organisations during the 1960s and 1970s did not necessarily usher in the supposed 'rolling back of the state' in the 1980s, but was instead indicative of a situation where voluntary organisations were viewed as an increasingly essential part of the response to a difficult and growing problem. A partnership between voluntary organisations and the state was being formed during the 1960s and 1970s, one which allowed for considerable flexibility on both sides.

## 3.1   Voluntary action and the state, 1960–1979

Before examining the nature of the relationship between drug voluntary organisations and the individuals and groups which surrounded them, it is necessary to outline some of the broader changes that occurred within voluntarism over this period, particularly the relationship between voluntary organisations and the state. The apparent upswing in voluntary action during the 1960s and 1970s has often been explained in relation to changes in the perception and nature of statutory welfare provision. Geoffrey Finlayson argued that it was the idea and practice of 'participation' that gave the voluntary sector a boost in this period, and while this was connected to concern with civil rights, it was also related to 'a certain waning of the initial optimism that the state alone would solve all of the problems of society.'[8] For Maria Brenton the growth of the voluntary sector in the 1960s and 1970s 'was as much a reaction of frustration to the deficiencies, size and inaccessibility of the state welfare apparatus as they were the result of pressures to participate and protest thrown up by a wider processes of social and cultural change.'[9] There was a feeling that statutory welfare was becoming remote, bureaucratic and unable to meet the needs of its citizens sufficiently.

At the same time, the state's view of voluntary organisations appeared to be altering. The Labour party had traditionally been opposed to voluntary action, seeing it as a tool of social control and as a partial solution that would be rendered obsolete by comprehensive state welfare services.[10] Yet, by the 1960s, an 'uneasy consensus' had begun to form around the political and practical value of voluntarism.[11] This was partly because of the issues already raised about the absence of participation in welfare services and their apparent inability to cater for social need comprehensively, but also because voluntary organisations had the potential to provide services at a lower cost. David Ennals, Labour Secretary of State for Social Services, described the use of voluntary effort in 1976 as 'pound for pound a better buy'.[12] The economic value of voluntarism was to attract even greater interest during the 1980s, but to some extent, the notion that voluntary provision might be a cost-effective solution to problems with service delivery in a time of fiscal crisis was already present.

A series of studies were commissioned throughout the 1960s and 1970s to investigate the contribution of voluntary organisations. Analysis of these reports suggests that as statutory interest in voluntarism grew, the relationship between the different sectors began to change. In 1966 the National Council for Social Service and the National Institute for Social Work Training established a committee under Geraldine Aves to enquire into the role of voluntary workers in social services.[13] According to Brenton, the major contribution of the Aves Committee was to focus attention on the benefits of closer integration of the voluntary and statutory sectors.[14] Crucially, however, volunteers were to work within existing statutory bodies rather than provide separate services. A somewhat different view was offered by the Wolfenden Committee in 1978. Appointed by the Joseph Rowntree Trust to review the role and function of voluntary organisations over the next 25 years, Wolfenden did not advocate radical change in the relationship between the voluntary sector and the state, but he did define the nature of this relationship more clearly. According to the committee, voluntary action should 'best be seen in terms of the ways in which it complements, supplements, extends and influences' statutory provision.[15]

In order to support the voluntary sector in its development of these roles, Wolfenden also recommended that more financial support be provided by central and local government. To some extent, this advice was taken up and statutory funding for the voluntary sector grew over the course of the 1970s. Central government grants to voluntary organisations in 1970–71 were estimated to be in the region of £2.5 million; by 1976–77 this had risen to £35.4 million.[16] The influx of government money to the voluntary sector did raise fears about the dependence of voluntary organisations on the state for financial support. Yet, as Brenton points out, 'Dependency does not rest all on one side, however, for this financial relationship also disguises the extent of increased dependence of government on the voluntary sector. The "partnership" that results is a means by which government can work out their own purposes through the agency of non-government bodies, just as much as a means whereby voluntary bodies secure their own futures.'[17]

Indeed, as a more detailed consideration of how this relationship worked in practice in the drugs field will demonstrate, the 'partnership' between the voluntary and statutory sectors was finely balanced, with both partners loosing and gaining by becoming more closely entwined. Looking at the interaction between drug voluntary organisations and the state, but also other actors including the medical profession, local communities and drug users themselves, provides a more rounded consideration of this relationship. The different roles performed by voluntary organisations in the drugs field could both complement and conflict with existing efforts to deal with the drug problem. Such a picture is further complicated by change over time. A large number of voluntary organisations in the drugs field were established in the late 1960s to meet needs not being addressed by the state. These organisations were often funded from charitable sources, but by the mid-1970s, many, if not most, of these organisations were receiving funding from the state. This shift suggests an increased interest from the state in the role that these groups could perform. And, as the decade closed this became much more that just a financial relationship: the state and the voluntary sector were actually working more closely together, as a brief consideration of the City Roads Crisis Intervention Centre will demonstrate. Such projects could be seen as paving the way for the rolling back of the state; however, this chapter will argue that there was actually a much more complex relationship at work here, characterised less by the retreat of the state than its relocation. The state was becoming more of a manager of welfare services than their direct provider.

## 3.2  Drugs, users and the local community

In the late 1960s a number of local organisations were created to meet the needs of drug users. A key group were the 'street agencies' offering advice, counselling and referral to other services or sometimes just 'a place to be' for drug users. Loosely speaking, these can be placed into four different categories: there were groups inspired by pastoral Christian theology; agencies that drew on notions of radical social work; groups that were inspired by the therapeutic community movement and others that grew out of community care and community psychiatry. Each of these categories and some example groups will be briefly considered before turning to a more detailed examination of one organisation, the Community Drug Project (CDP). Although no agency can be said to be 'typical' as these clearly differed quite considerably, many street agencies faced similar problems, and so the issues dealt with by the CDP exemplified some of the broader tensions around this kind of work.

As was noted in the Introduction, Christian organisations had a long tradition of working with drug users, particularly in providing residential care. But during the mid-1960s, a number of Christian groups and individuals became concerned about street drug use, often in connection with their work with homeless young people. A prominent figure was the Anglican priest, Kenneth Leech, who began his career in the early 1960s in the East End of London, where he encountered heroin use amongst some young people. In 1967, as a result of his experience of working with drug users, Leech was asked to go to Soho with a specific brief to see what the churches could do about drug use in the West End.[18] He became involved in the Soho Project, which had been established in 1966 by Barbara Ward, a professor of youth and community care at Goldsmiths College. Leech allowed the Soho Project to take over the basement of St Anne's Church (where he was assistant curate) and from there the project conducted what was described as 'outreach work.'[19] Leech, Ward and other volunteers at the Soho Project, spent time talking with young people in night clubs and bars about their drug use and other problems. Leech described this activity as pastoral care, but with strong elements of social work. He wrote, 'In practice there is considerable overlap, [between pastoral care and social work] since the priest is concerned with the whole person, with man in society orientated Godwards.'[20]

Around the same time, other, more secular, groups also began to approach drug use from a social work perspective. In 1964 Notting Hill Social Council appointed a social worker to look into the possibility of doing 'detached work' with young 'drifters' or homeless people. This gradually developed into the Blenheim Project, moving to a building on Portobello Road in 1969. The Blenheim Project regarded itself as a 'pre-treatment agency', giving homeless young people, and particularly those using drugs, a 'place to be'.[21] The project offered practical services such as washing facilities but also gave advice and aimed to achieve 'positive changes in the client (such that he or she is cleaner, pays rent, makes decisions etc.)'[22] Although the Blenheim Project was largely run by professional social workers, they adopted what they called a 'permissive approach', allowing the client to do what he or she wanted. However, Nicholas Dorn and Nigel South in their analysis of the Blenheim Project found that this 'radical social work' approach was not entirely successful, and by 1976 the project had abandoned its open-door policy in favour of an appointments system.[23]

Another agency which began as a drop-in centre was Lifeline, located in Manchester. Lifeline was one of the few non–London-based voluntary organisations of any type established in the 1970s. Lifeline grew out of a failed attempt by psychiatrist Dr Eugenie Cheesmond to establish a residential therapeutic community in the grounds of Park Side psychiatric hospital in Macclesfield. Cheesmond was able to obtain funds from Manchester Social Services to set up a day centre and Rowdy Yates, a former drug user himself, became involved in the project. Yates described as Lifeline in its early days as a 'therapeutic soup kitchen', offering food and a 'place to crash' for drug users, alongside the opportunity to take part in physical work repairing the building in which the project was based.[24] Lifeline referred people on to residential therapeutic communities, as well as developing some of their own programmes, such as the Bail Release Scheme where they took clients on bail for drug offences, gave them a programme of work and encouraged them to enter residential rehabilitation.[25]

The different inspirations and approaches behind the street agencies established in the late 1960s and early 1970s highlights the existence of a diverse and vibrant network of voluntary organisations around illegal drugs in this period. The presence of such a range of groups also points to a degree of uncertainty around drug use. Illegal drug use on any scale was a comparatively new problem and the absence of clear expertise around drugs meant that the field was relatively open to new ideas and organisations. Psychiatric expertise was beginning to develop around the treatment of addiction at the DDUs, but there was a feeling amongst many groups that statutory services for drug users were inadequate. A background note on the Community Drug Project by the sociologist Gerry Stimson, who was involved with the project in its early days, stated that, 'There is increasing concern among doctors and psychiatrists that conventional medical treatment may not be sufficient for curing drug addiction. Drug use is not only dependence on a drug but involves social ties, a personal image and way of life.'[26] The street agencies may have differed in their initial approach, but they were united by their desire to help drug users within the community.

A key example of the community-orientated approach can be found in the work of the Community Drug Project. The actual driving force behind the establishment of the CDP was Dr Griffith Edwards and his newly created Addiction Research Unit (ARU) at the Institute of Psychiatry. Edwards already had considerable experience of working with voluntary organisations. He was instrumental in the setting up of the Camberwell Council on Alcoholism in 1962, Giles House (later Helping Hand) in 1964 and Rathcoole House (later the Alcohol Recovery Project) in 1966.[27] Edwards also played a key role in the establishment of the Phoenix House therapeutic community for drug users (see Chapter 1) and was a firm believer in the value of voluntary action in the drug and alcohol field. The CDP, which began life as a part-time evening club in the crypt of St Giles Church in Camberwell in July 1968, and later moved to rooms above a greengrocer on a main street, also in Camberwell, was designed to have three functions.[28] First, to provide a place where the drug user could come and sit and talk, have a coffee and so on. Second, to make available certain social services such as help with finding accommodation. Finally, to help drug users develop relationships with non-drug users. These functions were to be directed towards the main aim of the CDP, which was to encourage users to enter treatment. The CDP was intended to 'establish an atmosphere where the addict can begin to see constructive alternatives to his drug use.'[29] The whole project was aimed at 'treating the problem of drug addiction at the community level.'[30]

'Community' was an essential concept to the work of the CDP, but this term seems to have had a variety of meanings, each with slightly different implications. Edwards's notion of 'community' appears to have been at least partly derived from ideas about community care and more specifically community psychiatry. Although the term 'community care' is itself imprecise and has changed over time, it has generally been used to describe a shift away from hospital-based treatment to care in the community.[31] From the late 1950s onwards there were concerted attempts to move long-term hospital patients, particularly the elderly and the mentally ill, into treatment in the community.[32] Douglas Bennett suggests that community psychiatry can be traced back to those who opposed the use of restraint in nineteenth-century asylums, but the steady reduction in the number of psychiatric inpatients following the Hospital Plan in 1962 undoubtedly resulted in an expansion of community psychiatry. Bennett also notes that this was 'a period of great confidence and enthusiasm in psychiatric services,' and some of this enthusiasm for community psychiatry was clearly present in Edwards's work.[33] Edwards stated in a letter to Captain John Brown, Secretary to the Trustees of the Attlee Memorial Foundation, and later Chair of CDP's management committee, 'We [the ARU] feel too that this is very much an experiment in *community* [his emphasis] psychiatry.'[34]

The CDP was obviously rooted in community care and community psychiatry, but in a sense it also drew on elements of what was being described as 'community medicine.' During the late 1960s public health was redefined as 'community medicine', within which the community physician would be responsible for community diagnosis and analysis of the health problems of the population.[35] The CDP seems to have seen drug taking within this context of community medicine, regarding drug use as a community problem requiring a community-based solution. In answer to the question 'what exactly do you mean by a Community Drug Addiction Project?' the CDP's suggested response was: 'It is our view that the problem [drug addiction] is very much a community one.'[36] As a result, the CDP saw itself as having a 'role in helping the community to see how to react to that community's drug problem in a way which is humane and constructive as well as self-protective.'[37] Indeed, another public health aspect of the CDP's work was to protect the wider community from the 'infection' of drug use: the CDP provided a room where users were allowed to inject drugs in order to prevent them from injecting in public toilets and telephone booths 'which was already causing a great deal of disquiet in the area.'[38]

The meaning of 'community' for the CDP had other, more internally focused, dimensions too. The CDP thought that one of the strengths of the project was that 'it is available – it is in the community. No appointments or formalities stand between the addict who wants help and that help.'[39] Once drug users started coming to the CDP, it was 'hoped that the addict would also be involved in the running of the centre and organising any social activities that might be wanted.'[40] This was designed to provide an alternative 'milieu', one which would allow users to 'become involved again in other social structures and indeed their own lives.'[41] There was clearly an element of user involvement built into the work of the CDP, as users were expected to create their own community. This was followed by an attempt to introduce more formal drug user involvement in the mid-1970s when an ex-addict worker was employed to work alongside professional staff and volunteers (including a social worker) at the CDP. This was not entirely successful as the individual started using drugs again and was asked to leave the project.[42]

This aborted attempt to employ an ex-addict was indicative of a some-what tense relationship between those running the CDP and the clients attending the centre. Involving users in the running of the project raised very real issues about who 'owned' the CDP, about its outlook and general purpose. Staff at the CDP became frustrated with 'an increasing number of attenders using CDP as a drug orientated centre as opposed to a social work support centre.' They felt the more 'positive' clients were being pushed out by those who wanted to use the centre as a place to 'crash', 'fix', 'deal' and 'score'.[43] Staff felt that they were in danger of 'colluding' with the 'drug lifestyle' and that the project was becoming a 'community dustbin'.[44]

A particular source of tension was over the injecting or 'fixing' room. The CDP provided a fixing room in order to prevent users from inject-ing in public, but it was something Edwards claimed he was always 'uneasy' with. Edwards was concerned that 'someone would over-dose and kill themselves one day, [and there were] a lot of drugs being handed around. I wasn't sure it was therapeutic in any way. There was no control over it.'[45] Staff at the CDP began to see the fixing room as 'being counter-productive to the possibilities of anything positive being achieved in the centre' and a source of continuing conflict as it 'increased the polarisation of feeling between "social worker" and "addict."'[46] In 1975 restrictions were placed on the use of the fixing room, with users only being allowed to inject on the premises at agreed times, but by 1976 injecting was banned altogether. Staff felt that it was 'impossible for the addicts to sustain any commitment to positive change in an atmosphere still dominated by the fixing room.'[47] In fact, the CDP was one of the last agencies to close its fixing room: both the Hungerford Project and the Blenheim Project had done so a few years previously.[48]

Alongside this level of user participation, the CDP also wanted to encourage the local community to become involved in the project. This stemmed from their view of drug use as a community problem requir-ing a community solution. Edwards wanted the local community to be 'truly involved in a helping and caring role', with volunteers working with professionals at the CDP.[49] However, reaction from the local com-munity to the project was not always so positive. A number of residents formed a protest group which complained that the CDP had 'thrust' the project upon a local community already experiencing a number of social problems. The Southwark Residents Group felt that the project had encouraged more users into the area, exposing children to drug use.[50] A public meeting to protest about the project was held, and coupled with 'uneasiness' from the 'local community about the presence of the CDP in the main street,' the project was forced to move to a disused Sunday school building in a less prominent location.[51]

While the CDP was quick to claim that its dealings with its immediate neighbours were good, and its relationship with local residents did improve, it was somewhat paradoxical that elements of the community appeared to be rejecting the notion of a Community Drug Project in their midst. Yet, it was this emphasis on community, no matter how contested, that marked out what the CDP was doing from the statutory services for drug users. As a more detailed analysis of the relationship between street agencies such as the CDP and the medical professionals concerned with treating drug use will demonstrate, voluntary organisations in this field were providing a different type of service, underpinned by a different way of viewing drug use and the problems this could cause.

### 3.3 The relationship between the street agencies and statutory services

As discussed in Chapter 1, the main statutory services for drug users during the 1960s and 1970s were the Drug Dependence Units (DDUs). The DDUs offered treatment in the form of either short-term withdrawal or long-term substitute prescription, known as 'maintenance', which was intended to prevent the development of a black market in illegal drugs. Yet, maintenance prescription became less common by the late 1970s as the DDUs moved patients onto time-limited prescriptions. This shift was indicative of what could be described as an increasingly 'medical' approach, with DDU staff tending to focus on 'curing' that patient by getting him or her off drugs.[52] Many of the voluntary organisations working in the drugs field contended that this approach was too narrow and that the DDUs focused on only the medical aspects of what was a much bigger problem. Release argued that, 'Drug dependency is usually the outward sign of a deeper personality problem. Unless the patient is given psychiatric and social support it is almost certain that he will return to drug use...Treatment Centres [DDUs] will not be able to operate effectively until social and medical care are fully integrated.'[53] Similarly, the CDP noted that with the establishment of the DDUs in 1968 'Britain chose to define the addict and his needs according to medical perceptions' yet 'with increasing experience it has become apparent that these medical perceptions are but minimally related to the contemporary needs of the people who attend the CDP.'[54] Workers at the CDP felt that, 'It could be said that a strictly medical response to the problem of drug dependency was failing to deal with crucial underlying emotional and social factors.'[55]

From these general criticisms more specific issues also arose. Staff working at the CDP believed that they were closer to their clients and knew how drug users viewed the DDUs. Susan Martin, a social worker at CDP commented that 'addicts do not feel that the clinics are treating them, but rather view the clinics as a steady source of drugs to which they are entitled.'[56] When this supply of drugs began to dry up in 1970s due to reduced maintenance prescriptions from DDUs, street agencies and other voluntary organisations working in the drugs field felt the effects. The CDP annual report from 1970–71 noted that 'the day centre now seems to be dealing with many more long term addicts who resent the pressure being applied to them, and see the less generous prescribing policies of some treatment clinics as a betrayal of their rights.'[57] Some voluntary groups were also critical of the DDUs exclusive focus on opiate drug users. Richard Paris, a worker at the Blenheim Project, asserted that, 'The Drug Dependency Units are coming in for mounting criticism on all fronts in their policy of seeing opiate-only dependents, usually after considerable delay and commonly supplementing their black market supply with oral methadone or tranquilisers.'[58]

The street agencies were well placed to observe changes in the drug scene, and from the early 1970s onwards, a number of groups expressed their concern about the growing use of barbiturates. The CDP contended that barbiturates were easy to obtain from GPs and taken by users as a substitute for heroin, particularly when this became less readily available from the DDUs.[59] Barbiturates posed specific problems for street agencies and other groups attempting to work with users. The CDP felt that barbiturate users were often aggressive or comatose; that barbiturates possessed greater potential for overdose; that coming off these drugs resulted in serious withdrawal symptoms; and that as barbiturates were not generally intended for injection they had to be prepared in such a way that was likely to lead to physical complications.[60] SCODA argued that, 'Existing agencies, particularly the drug treatment clinics must re-assess their function and adapt to meet these new and complex problems.'[61] An addict writing in *News Release* contended that the increase in barbiturate use was largely the fault of the clinics for failing to deal with opiate users effectively: 'more addicts die for barbs overdoses than anything else. I really think a lot of the blame for this must lie with the clinics.'[62] Partly as a result of the DDUs refusal to see barbiturate users, the street agencies claimed to be in contact with a different section of the drug using population. The co-ordinator of SCODA, Bob Searchfield, noted in 1976 that voluntary groups and casualty departments were encountering a group of drug takers 'for whom the present treatment facilities are inappropriate.' This was not, he suggested, necessarily a new problem but '[w]hat is clearly evident is the failure of treatment and rehabilitation services to effectively treat, withdraw and socially reintegrate the vast majority of drug addicts.'[63]

It is clear that many voluntary organisations were critical of the service provided by the DDUs, but some of the staff at the DDUs took an equally dim view of the actions of street agencies. DDU doctor Margaret Tripp told the CDP that some clinics believed that day centres such as CDP made the drug problem worse as they 'perpetuate a drug sub-culture.'[64] A questionnaire about the role of day centres sent by the CDP to DDUs revealed that attitudes towards street agencies were at best ambivalent. In answer to the question 'Does the existence of a day centre such as ours make the situation better or worse?' only one DDU answered 'better'; one answered 'better, but [it] depends on the balance between permissiveness and authoritarianism'; one answered 'probably better', but that day centres 'must be seen as an experimental trial.' Furthermore, two DDUs answered that day centres made the situation 'worse' because they 'perpetuate the junkie ethic.'[65]

However, other DDU doctors saw the relationship between street agencies and clinics in a more positive light. Dr James Willis, Consultant Psychiatrist at Guy's Hospital told Edwards that 'I tend to see the roles of the CDP and my clinic as being complementary in that we don't as it were try to do each others jobs in any way.'[66] For his part Edwards was 'very keen that the CDP should never accidentally acquire the image of being a project which was in some way a threat to the clinics.'[67] Of course, the relationship between the CDP and the DDUs was always more likely to be built on consensus rather than conflict, given Edwards's professional orientation as a psychiatrist. Moreover, DDU doctors, such as Martin Mitcheson and Margaret Tripp, sat on the project's management committee. Yet, other groups also appear to have been capable of viewing the relationship between voluntary organisations and statutory services in a similar way. SCODA noted in their 1976–77 annual report that, 'The voluntary sector, was, to a large extent, developed to complement the drug treatment units.' The difficulty, according to SCODA, was putting this into practice. That: 'Even where voluntary and statutory agencies are working on the same problem, co-operation and co-ordination barely exists.'[68] This is perhaps unsurprising given the level of antagonism between the DDUs and some of the street agencies, and also their different views on the nature of the drug problem. In fact, it could be argued that voluntary and statutory services did not necessarily need to work together if their activities largely complemented one another. By dealing with problems such as barbiturate use, the street agencies were actually fulfilling the role envisaged for them by the Wolfenden Committee on the Future of Voluntary Organisations. Street agencies and other voluntary groups were complementing, supplementing, extending and influencing statutory provision.

## 3.4    Voluntary organisations and research

Another area where the voluntary sector could be described as taking on a largely complementary role with respect to statutory effort was around research into the problems posed by drug use. When heroin addiction first began to cause popular and political concern during the late 1960s, little was known about the condition or how to treat it. Dr Thomas Bewley, a consultant psychiatrist at Tooting Bec Hospital, told an interviewer from the journal *Addiction* that he became an 'expert' on addiction when, in 1964, he had seen only 20 addict patients, but this was far more than any other doctor had seen. He commented, 'This was how I became an "expert." I knew little, but everyone else knew less.'[69] In 1967, the Addiction Research Unit (ARU) was established at the Institute of Psychiatry to fill this knowledge gap. The ARU was funded by the Medical Research Council and headed by Griffith Edwards.[70] Although the ARU was not itself a voluntary organisation, it was closely involved in the establishment and initial management of a number of voluntary projects including Phoenix House and the CDP. ARU staff, including Adele Kosviner, Martin Mitcheson and Gerry Stimson, played a key role in the early days of both CDP and Phoenix House. ARU researchers also conducted studies on the work and effectiveness of these organisations.[71] This close connection between research and voluntary activity stemmed from Edwards's belief that when trying to deal with the problems posed by drug use 'the sense of experiment and the attempt to research' were intertwined. 'Everything we did' he said 'was from a research staff, a research base.'[72]

Other organisations, not directly connected to Edwards or the ARU, also carried out their own research. For example, Release produced the *Release Report on Drug Offenders and the Law* in 1969 and periodically commissioned studies of their own activities and their impact. In part, this research could be seen as a way for voluntary groups to justify their own existence by pointing to a clear need not being met by statutory services and, therefore, the importance of their own work, but research into drug issues was also taken on by a dedicated voluntary organisation, the Institute for the Study of Drug Dependence (ISDD). The ISDD was the brainchild of Frank Logan, a former Home Office and United Nations Division of Narcotics official. In 1967, Logan had been trying to write an article about drug use and had found that not only was there a lack of factual information on the topic, but there was also no obvious library or collection to turn to for help in locating relevant material.[73] Logan decided to create an institution where individuals could gain access to published work on drugs and also 'to provide a centre where the subject [drug use] is under continuous study.'[74] Jasper Woodcock, who began work at the ISDD in 1969 as an information officer and later became the Institute's director, stated that there was a perceived need for neutral, objective information about drug use. ISDD was therefore set up on the basis that the information it provided would be without bias; that it would be believable by everybody.[75] The Institute was intended to be 'non-political and entirely objective in its approach. While aiming to co-operate with Government Departments... it would be free to express its own views.'[76] Logan thought that ISDD's independence could be guaranteed by its 'non-governmental' or 'non-official' status: that there was a particular need for a voluntary, as opposed to statutory, body in this area. To some degree, it would appear that state concurred. Sir Harold Himsworth, deputy chairman of the Medical Research Council remarked in 1968 that 'I personally think that there is a great deal to be said for a voluntary body in a field concerned with [an] illness in which suspicion on the part of the patients is the characteristic feature. These views are not only my own. I have checked with Godber [the Chief Medical Officer]... and they are his also.'[77]

The ISDD opened on 1 April 1968, and was initially based in the attic of the Royal Society's Chandos House, near Oxford Circus, moving to a house in West Hampstead in 1973.[78] A reference library of material on drug use was assembled and catalogued, a task begun by volunteer amateurs and later continued by professional library and information staff. The Institute also developed its research into drugs, setting up a research unit in 1972 which employed the organisations' first full-time staff.[79] It was through this research that the ISDD had the potential to influence government policy on drugs, despite their stated aim to be neutral and objective. Woodcock stated that he felt he had often wanted the ISDD to get more involved in policy issues.[80] One such area was over cannabis. In 1973 the ISDD established a working party to examine various options for controlling cannabis use.[81] Woodcock noted that while the report avoided recommending any particular form of control, it did skirt close to the policy arena and 'quite a few [ISDD] council members were unhappy about it.'[82]

## 3.5    The relationship between voluntary organisations and local and national government

ISDD's flirtation with a role in policy draws attention to the interaction between drug voluntary organisations and local and national government. This relationship was clearly multi-faceted and needs to be examined from a number of perspectives. The first concerns the funding of voluntary organisations by statutory bodies. In 1978 the Wolfenden Committee recommended that more financial support be provided by central and local government to voluntary organisations working in all fields.[83] In health, Section 64 of the Health Services and Public Health Act (1968) permitted the DHSS to give grants to voluntary bodies working on a national basis, and local authorities were expected to support local projects.[84] The Voluntary Services Unit (VSU), established in 1973 and based in the Home Office, also occasionally gave grants to groups working in the drugs field, such as Release (see Chapter 2).

Despite this expansion in statutory support for voluntary organisations many groups, such as the CDP, were established without statutory support. The project had initially been funded by the Attlee Foundation, a charitable trust. The Attlee Foundation supported the project because 'the Trustees believe that it is their duty to use the funds which are available to the them … to work in fields which are complementary to government and national policy, but for which no government funds are available.'[85] The CDP later received donations from the Cadbury Foundation and the City Parochial fund.[86] But, by the early 1970s, the CDP's costs could not be met by charitable sources of income alone, and there was a feeling amongst the project's staff and trustees that as the CDP had proven its worth it 'should now be financed by public funds.'[87] The CDP thus approached a series of statutory bodies. The Home Office were 'in no doubt about the value of the project' but concluded that 'the object of the Community Drug Project lies primarily in the medico-social field' and so were unable to support the project.[88] Dr Sippert of the Department of Health and Social Security (DHSS) also told representatives from the CDP in 1971 that the DHSS 'had high regard for the work of the CDP and was most anxious that it should continue operating' but the Department was reluctant to fund the organisation as it operated at a local level.[89] The DHSS therefore suggested that the London Boroughs

Association (LBA) was the appropriate agency to approach for statutory support.[90] The LBA was initially reluctant to fund the CDP and other drug voluntary organisations as it felt it was beyond its resources, but in 1972 the Association gave the CDP £12,000, as well as finding money to support the Hungerford Project and a number of residential facilities, including Phoenix House.[91] The LBA continued to support the CDP, but even these funds became insufficient, and in 1982 the DHSS agreed to give money to the CDP and two other street agencies to prevent them from closing.[92]

This central support for a local project despite a stated intention not to fund such groups attests to the value which the government placed on the work of the CDP and organisations like it. Indeed, as was seen in Chapter 2, such funding for specific voluntary organisations was not that uncommon: Release received a grant from the Home Office through the VSU. This support was in line with the general increase in central funding for voluntary organisations in all areas, and the drugs field was clearly no exception. SCODA estimated that in 1975, £700,000 had been awarded to voluntary groups working in the drugs field.[93] Two key organisations in receipt of central government funds were SCODA itself and the ISDD. In its early days, the ISDD was reluctant to accept statutory funding as it felt it might compromise its reputation for independence. As a result, the Institute was initially funded from charitable sources, including the Harry Stewart Trust and the US-based Drug Abuse Council.[94] But, in 1972 the DHSS offered the ISDD a grant of £10,000 over three years to enable the organisation to develop the library and employ professional staff. Woodcock argued that 'ISDD by then felt sufficiently secure in its independence to accept.'[95] When the ISDD faced closure, a few years later, as a result of financial crisis the DHSS stepped in, and by 1977 the government department was providing core deficit funding, supporting the ISDD in its key activities.[96] However, Woodcock was quick to point out that the ISDD was only receiving between 40 and 50 per cent of its income from the DHSS, a situation, he maintained, that allowed the Institute to retain a considerable degree of independence.[97] The former ISDD director recalled a couple of occasions when the DHSS had attempted to influence how the organisation allocated its resources, but in each case he stated that he was able to tell the department that 'none of our activities are solely funded by you … and so you have absolutely no business telling us how to allocate the money.' Woodcock said that ISDD was able to tell the DHSS to '[g]o away basically, and both times they went away.'[98]

This remarkable degree of tolerance on the part of the state can perhaps be explained by the importance the government placed on the ISDD's work. A letter from a DHSS official to a counterpart in the Department of Education and Science observed that the 'ISDD is considered by those working in the field of drug abuse to be fulfilling a useful role.'[99] Another official recorded after a visit to the ISDD that he was 'very impressed with the quality of the Institute's achievements.'[100] Woodcock himself felt that the DHSS had allowed ISDD to retain its independence because 'they must have thought there was some value' in their work.[101] It would seem that the DHSS was prepared to allow the ISDD to retain its freedom as the organisation was already fulfilling a role which the state valued. This specific example is supported by what seems to have been a more general trend. Brenton, writing in 1985, argued that, 'It stands to reason that they [government controls on voluntary organisations] will not be excessive, given the importance to government of the voluntary bodies freedom from red tape and its flexibility.' Moreover, she continued, 'a common sense judgement would also suggest that for government departments to monitor very closely the activities of every one of the hundreds of agencies subsidised would be both impracticable and almost impossible.'[102]

Yet, the government did not deal with all voluntary organisations in the drugs field in the same way. A somewhat different relationship with the state can be observed in the experiences of another national drug voluntary organisation: the Standing Conference on Drug Abuse (SCODA) which grew out of a series of meetings held by voluntary organisations working in the drugs field in the early 1970s. Concern was expressed about the 'lack of communication between the various bodies' and there was a perceived need to 'effect closer co-operation in their work.'[103] SCODA was created in 1972 to co-ordinate the activities of drug voluntary organisations. Its aims were; first, to provide a forum where information on the pattern of drug use can be shared and where policy recommendations can be formulated; second, to identify needs and assist in developing services; and finally to encourage research into drug use, evaluate services and stimulate the training of agency staff.[104] To fulfil these aims SCODA carried out a range of activities. It took on a considerable degree of policy and campaigning work (which will be discussed in detail later) but also provided services to drug voluntary organisations who became members by purchasing a 'share' in SCODA for £1.[105] SCODA held regular meetings with groups of voluntary organisations working in different areas of the drugs field, such as the therapeutic communities and the street agencies. SCODA staff

went out to visit agencies in action and offer advice on their organisation and running. SCODA produced a number of publications, including a members' newsletter and a guide to treatment and rehabilitation
facilities.[106] The organisation also became involved in the development
of training programmes for drug voluntary organisation workers, and
held conferences to bring the various organisations working with drug
users together.[107]

Alongside the services SCODA offered to member agencies were
the services it offered to the government. SCODA saw one of its main
functions as being to 'bridge the communications and information
gap between street level agencies and government ministries.'[108] The
potential value of this role was clearly recognised by the state: in 1972,
SCODA received a three-year grant from the DHSS to fund its work.[109]
A SCODA worker from this period noted that, 'Government always
likes to have a mechanism for communicating to a wider audience
rather than [to] have to deal individually with every single body.'[110]
Indeed, there was a more general policy shift towards the creation of
co-ordinating bodies for the voluntary sector. In 1978 the Wolfenden
committee examined the role played by intermediary bodies at the
local and national level in co-ordinating voluntary action and recommended that more of these groups be established. Wolfenden suggested that these bodies should have functions very similar to those
of SCODA, such as identifying areas of need, offering services to
other organisations and representing the views of these groups to the
statutory sector. Moreover, Wolfenden also recommended that these
organisations be funded by the government, as they would have little
charitable appeal.[111]

SCODA clearly fits into this model in terms of its functions and in
terms of its level of statutory support. Throughout the 1970s the DHSS
provided SCODA with its sole source of income, so that by the end
of the decade SCODA was in receipt of over £38,000 a year.[112] The
DHSS justified this financial assistance (for which they had to have
special authority from the Treasury) as SCODA operated 'in a field
unattractive to the public and if unsupported [SCODA] would have to
compete for funds with its member agencies whose direct service has
more immediate appeal.'[113] Attached to this funding, however, was
an increasing number of conditions. In 1976, the DHSS decided to
review its support for voluntary organisations working in the drugs
field and asked SCODA to submit a report on their current work and its
future directions.[114] Although the DHSS decided to continue to fund
SCODA, an official posed the question as to whether their grant to

the organisation 'should have more conditions [attached to it] more relevant to current views and policies.' While these conditions were not necessarily at odds with SCODA's work – the civil servant wanted to see SCODA 'help inform us of places where statutory authorities are not as involved as they might be' – the possibility that the DHSS might become more directive raised questions about SCODA's ability to remain independent.[115]

Yet, when SCODA's policy and campaigning work during the 1970s is examined, there seems to have been no obvious attempt by the DHSS to influence SCODA's work. A SCODA worker stated that 'No assistant secretary at the DoH [Department of Health] ever found me quiet and calm to follow what I was told to do if I thought they were wrong.'[116] From the outset observers from the DHSS and the Home Office, as well as the LBA, sat on SCODA's management committee, but these individuals do not appear to have exerted any influence over SCODA's work. Indeed, SCODA was at times openly critical of the government's policy on illegal drugs and of statutory service provision. A SCODA worker asserted that 'it was perfectly reasonable to receive funding [from the state] but also to criticise [the government].' The voluntary sector, he argued, was the 'loyal opposition.'[117] SCODA were, for example, able to point to the weaknesses in statutory services as they saw them, and also campaign successfully against additional controls being placed on barbiturates.[118]

SCODA's close relationship with the state may actually have helped the organisation to become more effective in its campaign work. SCODA staff noted in the organisation's 1974–75 annual report that, 'From the beginning SCODA has enjoyed the interest and support of various government departments and this relationship is highly valued by SCODA member agencies who now feel they have a greater access to public policy making bodies.'[119] Taking money from the government also helped to enhance SCODA's image and allowed the organisation access to the policy community that was beginning to form around illegal drugs. SCODA's first co-ordinator, Bob Searchfield, sat on the ACMD, and he was followed by David Turner when he took over in 1977. As will be seen in Chapter 4, SCODA played a key role in bringing the ACMD's attention to the growing drug problem in the late 1970s, and in campaigning for the introduction of central funding for drug services in the early 1980s.

But, for voluntary organisations such as SCODA, dealing with statutory organisations could prove to be a frustrating experience. A SCODA worker felt that during the 1970s 'there really was, in terms of policy, of having a drugs policy or any specific responses ... there was absolutely no sense of urgency.' SCODA therefore developed a number of alternative strategies for influencing drug policy. The SCODA worker maintained that 'if you want to influence and adjust and increase awareness conversations, quiet talks [and] quiet presentations' were much better than 'shouting'. This was because 'if you shout all the time, all people do is put their ear plugs in. Only shout when it's absolutely essential that they hear.' The SCODA official stated that he was 'not keen on the loud noises when you could do much better by other mechanisms.'[120] These other mechanisms included the creation of the All Party Parliamentary Committee on Drug Misuse in 1983, lobbying Conservative back benchers to ensure that drugs did not fall off the agenda and working with all political parties so that no major differences should emerge in their respective drug policies.[121] Yet, such lobbying tactics were not always successful. SCODA was unable to defend the removal of ring-fenced funds for drug services in the early 1990s, and its reliance on Department of Health funding in the mid-1990s resulted in significant changes in the organisation, including the resignation of Turner in 1994.[122] However, during the 1970s SCODA does not appear to have been inhibited by its close financial relationship with the state from taking controversial action. Like voluntary organisations in the drugs field more widely, SCODA complemented, supplemented and extended existing statutory provision. But, there were signs towards the end of the 1970s that this relationship was shifting once more, with the state and the voluntary sector increasingly working together as well as in complementary roles.

## 3.6 Combining voluntary and statutory approaches

This ability of the drugs voluntary sector to work with the state as well as outside it can be observed in the establishment of the City Roads Crisis Intervention Centre. Since SCODA's inception, voluntary organisations in London had been campaigning for the creation of a short-stay residential facility to meet the needs of multiple drug users.[123] SCODA was concerned about the rising number of drug users taking a range of drugs including opiates and barbiturates. Statutory services catered almost exclusively to opiate users, leaving a large number of young, multiple drug users without appropriate care.[124] At the same time, NHS doctors were noticing that barbiturate users were increasingly presenting at Accident and Emergency Departments (A&E). Dr Hamid Ghodse from the ARU conducted a study of London hospitals and found that barbiturate users often overdosed and the lack of available facilities meant they had little choice but to go to A&E, where they caused disruption to other patients.[125] Ghodse suggested that once barbiturate overdose patients had been detoxified 'there ought to be some immediate form of social intervention.'[126] The City Roads crisis intervention project thus grew out of the concerns of both statutory and voluntary workers.

Combining the efforts of the voluntary and statutory sectors was at the heart of the City Roads project. The crisis intervention centre was intended to be an example of 'cooperation between several statutory bodies as well as voluntary,' but the establishment of the City Roads was protracted precisely because it involved multi-agency working.[127] The project took seven years to come to fruition, with the centre finally opening in April, 1978. This was partly due to a series of lengthy negotiations between SCODA, the DHSS and the LBA over the funding of City Roads.[128] The DHSS eventually agreed to meet a substantial portion of the costs of City Roads as 'the unit will provide a unique opportunity to assess whether voluntary effort can help significantly drug misusers at a critical point in their career i.e. when discharged from accident and emergency departments after overdosing.'[129] City Roads was designed to be a short-stay residential unit offering what was termed 'crisis intervention' to young, multiple drug users in the London area. 'Crisis intervention' meant providing a 'warm, sheltered environment where nursing, medical, social and psychiatric support is readily available.'[130] City Roads was not intended to offer a complete programme of treatment and rehabilitation, but rather to stabilise drug users and refer them on for further treatment.

The amalgamation of statutory and voluntary approaches can also be seen in the staffing and management of City Roads. SCODA initially played a key role in managing the project, but when City Roads finally opened it became an independent voluntary organisation. The centre was staffed by multi-disciplinary teams made up of a social worker, a nurse and a care assistant, with the idea that the different workers would learn from each other.[131] This joint working was seen as being essential to the experimental nature of City Roads. Furthermore, because City Roads was an untried project, funded in large part by the state, an evaluation of its work was built in from the start, a new development which was to become more common as the controls on statutory funding for voluntary bodies increased.[132] A team of researchers from Birkbeck College made up of Anne Jamieson, Alan Glanz and Susanne

MacGregor were funded by the DHSS to evaluate the project. Jamieson and her colleagues found that City Roads was 'a voluntary project but it was not allowed total freedom in the way it could operate, being supervised in its early days by the DHSS and continually by a management committee increasingly dominated by the statutory sector.' They concluded that 'City Roads might then be seen as a hybrid, an experiment in cross-fertilisation.'[133]

## 3.7   Conclusion

The experiences of City Roads demonstrated that not only could the voluntary and statutory sectors work together, but there might also be a number of advantages by doing so. Both parties stood to gain by developing closer connections. Drawing closer to the state allowed voluntary organisations access to considerable sources of financial support, vital in a field where public sympathy, and so charitable donations, was in short supply. At the same time, accepting statutory funding enhanced the reputation of voluntary groups such as SCODA and gave them a way into the policy community around drugs. In SCODA, the state gained a useful co-ordinating body, an organisation which could represent the views of the voluntary sector around drugs without requiring officials to deal individually with the burgeoning number of local voluntary groups providing services to drug users. Many of these voluntary groups, such as the CDP, emerged to fill a perceived gap in statutory services. This also suited the state, as though drug use was regarded as a growing problem, statutory services alone (in the form of the DDUs) were clearly inadequate. As Brenton has noted government, in its desire to 'do something' about a particular problem often turned to the voluntary sector as it valued the flexibility and innovative nature of voluntary action. Voluntary bodies, she contends 'add an instant task force capability to the cumbersome machinery of government in a way that rapidly accumulates instant political capital for relatively small financial outlays, particularly for policies which governments are not keen on.'[134]

Gaining monetary support from the state for their activities did raise a number of issues for voluntary organisations, but most felt able to preserve their independence. The state intervened infrequently in the activities of the voluntary bodies funded, as these organisations were already performing a role that the state valued. However, as statutory support for voluntary effort increased still further during the 1980s, so too did the perceived need to account for this spending. Voluntary organisations funded by the state had to provide value for money and to make certain that their activity did not contradict statutory policies and goals. Chapter 4 will explore how the changing view of the state's role in welfare provision and increased funding for voluntary organisations during the 1980s could alter the nature of voluntary-statutory relationship still further, as the state appeared to be 'rolled back'.

# Part II
# 1980s

# 4
# Rolling Back the State?
# The Central Funding Initiative
# for Drug Services

The partnership formed between voluntary organisations and the state during the 1960s and 1970s around the response to drug use underwent significant change in the 1980s. On 1 December 1982 the Secretary of State for Social Services, Norman Fowler, announced the establishment of the Central Funding Initiative (CFI) for drug services. The CFI offered both local authorities and voluntary organisations the opportunity to bid for grants to allow them to provide services for drug users in England.[1] Similar schemes were to be set up in Scotland and Wales.[2] Initially, the CFI was designed to provide £6 million over three years, but the programme was extended in January 1986, partly in response to the discovery of HIV/AIDS amongst injecting drug users. Under the initiative, a total of £17.5 million pounds was awarded to statutory and voluntary organisations between 1983 and 1989.[3] Ministers hoped that 'authorities and voluntary bodies will respond imaginatively to this initiative.'[4] Consequently, a range of projects were to be encouraged, 'including innovative experiments.'[5]

This apparent need for innovation and a dedicated stream of financial support for drug services was prompted by changes in the pattern of illegal drug use in Britain. The 1980s saw a significant rise in drug use, and the use of heroin in particular. The number of known heroin addicts increased by more than 10,000 over the ten years from 1978 to 1988, a six-fold increase.[6] Heroin use was also increasingly to be found in urban areas across the country, and not, as it was previously thought, just confined to London.[7] In 1985 the Conservative government asserted that 'the misuse of drugs is one of the most worrying problems facing our society today.'[8] Growing fears about drug use prompted a flurry of activity from both central and local government, from law enforcement bodies, voluntary organisations and health professionals.

The CFI was a key part of this response, but the initiative was more than just a reaction to the spread of drug use to the regions: it was indicative of broader changes within the drugs field and within health and social policy more generally. The CFI was designed to foster a multi-disciplinary approach to drug use, providing a range of services, such as residential rehabilitation and street-based counselling, to drug users. This was in contrast to the primarily medically-orientated response to drug use in existence since the late 1960s, based around out-patient treatment in the DDUs.[9] Through the CFI, less emphasis was placed on treatment alone, suggesting greater attention to the social as well as the medical consequences of drug use. This was to be achieved by involving a wider range of agencies in providing services for drug users. Particular encouragement was given to voluntary organisations as these were regarded as being more flexible than statutory bodies, and thus better equipped to respond in new ways to the rapidly developing drug problem.

Using the voluntary sector in this way was, of course, not an entirely novel development, but the extent to which the state came to rely on non-statutory organisations in the 1980s needs to be understood in the context of the Thatcher government's attempts to 'roll back the state'. Voluntary organisations were encouraged to play a greater role in health and welfare services in this period as the state supposedly retreated from direct service provision. By the end of the decade, private, statutory and voluntary bodies were expected to bid for contracts to provide specific services within an increasingly mixed economy of welfare. However, there was a paradox in these changes, as social policy commentators were quick to point out: increased statutory funding of the voluntary sector tied this more closely to the state than before, a development which not only raised questions about the independence of the voluntary sector, but also about the extent to which the state had really reduced its role in welfare.[10] By examining the CFI for drug services in detail, this chapter will, therefore, highlight specific tensions inside the drugs field and also within broader moves to 'roll back the state'.

## 4.1   A developing problem: Drug use in the late 1970s–1980s

During the 1980s illegal drugs came to dominate popular and political agendas as never before. This was due (at least in part) to a significant change in the size and scale of drug use. Although accurate figures for drug use are notoriously difficult to obtain, some indication of the number of drug users can be gleaned from the Home Office Addicts Index. Established in 1934, it recorded the names of addicts known to the Home Office, based on information derived from police inspections of pharmacists' records and notifications of addiction by doctors.[11] There had been a steady increase in the number of known addicts since the early 1960s, but the total number of cases largely remained under 2,000 until the late 1970s. From 1977 onwards the number of known addicts rose from 2,016 in 1977 to 2,402 in 1978 and 2,666 in 1979. This trend persisted into the 1980s: between 1980 and 1981 the number of notified addicts increased by almost a thousand, a 44 per cent increase.[12] Notifications to the Home Office continued to rise throughout the decade so that by 1987 there were 10,389 known addicts.[13] Despite better reporting of addiction, these figures were thought to be a significant underestimate.[14] The 'real' number of drug users was much debated and various multipliers based on epidemiological indicators were brought into play. Some commentators argued that it was quite possible that there were as many as 100,000 heroin addicts in Britain by the end of the 1980s.[15]

In addition to a numerical increase, drug use appeared to be spreading across the country. Before this period the 'British drug problem' could perhaps more accurately be described as the 'London drug problem'. By the end of the 1970s it was clear that this was no longer the case. Lord Denbigh, in a speech to the House of Lords on drug addiction in 1979, stated that over the past seven years there had been a 127 per cent increase in the number of known addicts residing outside the London area.[16] Between 1978 and 1981 most regions saw incidences of addiction double, and a number of studies identified significant heroin use in places such as Merseyside, Manchester and Glasgow.[17] For the authors of one of those studies it was clear that this was a 'new phenomenon'.[18]

This new phenomenon has usually been explained in terms of supply and demand. During the 1960s and early 1970s most of the heroin used by addicts was legally produced and prescribed, if illegally traded, between users. As the drug could be obtained on prescription from licensed medical practitioners, there was virtually no black market in heroin. This changed dramatically in the late 1970s. Most commentators point to an influx of heroin into Britain following the Iranian Revolution in 1979.[19] The amount of illicit heroin seized by police and customs, likely to represent a fraction of the total smuggled into the UK, rose considerably. The authorities seized just 3 kilograms of heroin in 1973, compared to 93 kilograms in 1981.[20] What is more, in real terms, the price of black market heroin fell by as much as 25 per cent between 1980 and 1983.[21] Significantly, most of this heroin was so-called 'brown' heroin, particularly well suited to smoking rather than injecting. This was important, as studies suggested that potential users who might be put off heroin by the danger and stigma attached to injection would feel less reluctant to try the drug if they could smoke it. By the mid-1980s it was clear that there was a plentiful supply of relatively cheap, illegally produced and distributed heroin to be found in Britain.[22]

Explaining the corresponding rise in demand is much more difficult and remains a sensitive and complex issue. Contemporary interpretations have been characterised by the sociologist Susanne MacGregor as being either of the 'Right' or the 'Left'. Those on the Right usually emphasised the immorality of young drug users and their further corruption by evil drug 'pushers'. In contrast, those on the Left argued that there was a link between drug use and growing deprivation and unemployment.[23] This was consistently denied by the Conservative government: the Home Office minister in charge of drug policy, David Mellor, told the House of Commons in 1986 that 'no link between the taking of drugs and unemployment has been established. The taking of drugs' he continued 'is much more closely linked to pressure from friends and curiosity.'[24] It is difficult to deny, however, that the 1980s were a time of considerable social, political and economic turmoil in Britain. Unemployment was high throughout the decade, reaching over 3 million in 1982. Furthermore, unemployment was particularly common amongst young people in urban areas hit hard by the re-structuring of the economy such as the North East, North West and industrial areas of Scotland – the very same regions that began to experience extensive drug use for the first time.[25] A possible connection between drug use and deprivation has become less controversial in recent years, with research in the US, Britain and other parts of Europe pointing to a strong association between drug use and the social environment, but the wider debate on the causation of drug use continues.[26] Indeed, caution must be exercised when making links between drugs and deprivation, as drug use in this period developed into a much wider social phenomenon. While drug use could undoubtedly be found in areas of high unemployment, it also could be found in areas of relative affluence. Journalist Marek Kohn detailed media responses to drug use in the 1980s and found that it was often portrayed as a problem for people 'who live on big estates – country ones and council ones.'[27]

Irrespective of the cause of the rise in drug use, what was beyond doubt was that existing services were unable to cope with the increased number of drug users and their widening geographical spread. Both DDUs and voluntary services were stretched thinly and unevenly throughout the country. During the late 1970s and early 1980s most DDUs were to be found in the Thames area; so too were the majority of advice and counselling services and residential rehabilitation facilities.[28] In these regions, services were often limited to a handful of interested psychiatrists and general practitioners. A Sheffield based psychiatrist stated in an interview that 'in the seventies our nearest, if you like, tertiary [specialist] unit was in Nottingham. And in the time I worked in Sheffield I think I managed to get two people admitted there from Sheffield.'[29] Even in London, services were increasingly overburdened by the growing numbers of addicts seeking treatment. Waiting lists for treatment at clinics lengthened: many addicts had to wait more than six weeks for a first assessment appointment at a DDU.[30]

Furthermore, as voluntary organisations had been arguing for some time, (see Chapter 3) there was evidence to suggest that the DDUs were encountering a declining proportion of drug users. A change in treatment policy had occurred in the 1970s replacing maintenance (long-term prescription of opiate drugs) policies with those based on abstinence: it was this development and the forces behind it which were also an important dynamic for the CFI.[31] From 1968 onwards, heroin addiction had been a formally notifiable condition: doctors were required to notify the Home Office when they came into contact with an addict patient, although many had been notifying informally for years before that.[32] In 1970, 46 per cent of notifications of heroin addiction came from DDUs, 48 per cent from prison medical officers and just 6 per cent from General Practitioners (GPs). By 1981 the proportion of notifications from DDUs had fallen to 36 per cent and notifications by prison medical officers to 16 per cent, whereas notifications by GPs had risen to 48 per cent of the total.[33] It would seem that drug users were increasingly deserting the DDUs.

In part this was because of changing patterns of drug use. There was a move away from the exclusive use of heroin and towards more 'poly-drug' use, with users taking a range of drugs including heroin, opiate substitutes and barbiturates. Yet, at DDUs, according to *The Lancet*, there was 'a near total preoccupation with opiate dependence.'[34] A major reason was also that, by the 1980s, the main treatment offered by DDUs was a rapidly reducing course of oral methadone, with an ultimate focus on abstinence, something many addicts did not find acceptable. This change in policy came, not from government, which took little interest in drug policy at the political level, but from the group of psychiatrists who ran the London DDUs, who had adopted a consensual policy in order to prevent the 'silting up' of the clinics with long-term maintenance cases and also as a means of regulating supply.[35] For these reasons, drug services established in the late 1960s were clearly not able to cope with the changing nature of the drug problem in the 1980s.

## 4.2   Recognising the problem: Voluntary organisations and the drug 'policy community'

Throughout the 1970s, drugs had not been a political issue or even a matter of great political interest; the response was a matter of consensus rather than one of party political division. And, it was not politics which eventually put the matter on the policy agenda, but rather the new forces emergent in drug policy in the 1980s. During this period a new 'policy community' began to form within the drugs field. There was a shift from a primarily medically orientated policy community to one that included voluntary organisations, researchers, civil servants and politicians.[36] The voluntary sector played a particularly important role within this group, as it was voluntary organisations which initially raised questions about treatment policy, principally through SCODA.

From the late 1970s onwards, SCODA began to report increasing contact with people from all over the country who were seeking help with drug problems but were unable to find local services.[37] SCODA was also critical of official statistics on drug use, arguing that the problem was much bigger than the government realised as: 'The experience of many voluntary organisations is that a high proportion of their client group remains un-notified and in consequence, officially "unknown".'[38] By the early 1980s, SCODA were reporting that while the number of drug users continued to grow, services had failed to expand resulting in additional pressure on existing facilities. This led the organisation to call for intervention from central government, arguing in their annual report in 1981 that

> Ad hoc responses at a local level are inadequate at a time when all the indices of drug problems show a rapid rise and when there is a dearth of specialist services in many parts of the country. Without a central government initiative which includes central financing, there can

be little hope that a suitable pattern and range of services can be developed.[39]

SCODA were not the only body calling for greater involvement from central government. The role of the Advisory Council on the Misuse of Drugs (ACMD) was crucial. The Council, originally convened in 1966 as the Advisory Committee on Drug Dependence, was established in 1971 as part of the changes in policy enshrined in the Misuse of Drugs Act. It was representative of a new style of expert committee set up in the 1970s in the health field and was to advise the government 'on measures to prevent and deal with the social problems arising from the misuse of drugs.'[40] Its membership also illustrated the widening 'policy community' round the drugs issue, with individuals from a wide range of fields connected with drug use such as psychiatrists, nurses, social workers, probation officers and voluntary agencies. The expert committee was an important instigator of new approaches in the drugs field, in particular through its influential report on Treatment and Rehabilitation, which was published in 1982 after a long gestation.

In 1975 the ACMD had created a working group to review treatment and rehabilitation services for drug users. The working group's interim report, published in 1977, expressed some concern that the number of drug users appeared to be rising and that treatment services might not be equipped to cope with such an increase, but the group felt that 'much can be achieved towards the establishment, at little or no additional cost, of the necessary facilities for the treatment of drug-misusers by the wise deployment of existing resources.'[41] This assertion did not convince everyone. A civil servant asked to comment on the resource implications of the interim report remarked that 'it will be a difficult task to persuade health authorities to provide additional resources or indeed maintain the present level, in the current economic climate, for this small group of patients.' He suspected that 'you will probably have to offer the carrot of special funding to achieve anything new.'[42]

This was a prescient statement. By 1979 it was increasingly clear to the ACMD that more resources would be needed to fund drug services. Once more, pressure came from the voluntary sector. David Turner, Coordinator of SCODA and a member of the ACMD, told the Council that the burgeoning market in illicit drugs and the growth of drug use outside London meant that 'treatment services must be expanded to cope with the impending explosion of demand.' For Turner it was essential that the ACMD's final report 'should explore areas of funding in relation to providing new services and to improving the options available to existing ones.'[43] Turner's was not a lone voice. Dr Anthony Thorley, a consultant psychiatrist in Newcastle, felt that 'without central funding the grass roots services would always live a hand to mouth existence as drug misusers were a stigmatised section of society.'[44] This view was reflected in the final report, *Treatment and Rehabilitation*, published in 1982. The ACMD noted that the stigma attached to drug addiction made funding services for drug users a low priority when allocating resources. To combat this, the report recommended that 'there should be increased funding, direct from central government, possibly by way of pump-priming grants.'[45] This, it was hoped, would lead to the development of multi-disciplinary services for drug users in every region.

The key facilitator of the new approach within government was, ironically, a doctor. Dr Dorothy Black, Senior Medical Officer at the DHSS with responsibility for drug policy, was a central figure in the stimulation and implementation of the new policy.[46] Black began work at the DHSS in 1981, but had previously been a psychiatrist involved in the treatment of drug users in a Northern city. She brought with her to the DHSS the knowledge that the drug problem had already spread beyond London, and also a sense that local authorities had little interest in funding services for drug users. During her time in the North she had also set up a local council on drugs and had worked with the voluntary sector.[47] Black's appointment was itself a tacit recognition that drug use was becoming a problem. Previously, drugs and alcohol had been dealt with together by the DHSS, with the greater emphasis placed on alcohol policy. A senior civil servant commented that by 1982 'people were beginning to pay more attention, partly because it was felt that there was an increasing number of people using drugs, which was supported by the figures of the Home Office Index, and so it was decided that something must be done.'[48] Behind this rational justification of change also lay a two-pronged strategy which sought to undermine the power of the London psychiatrists. It aimed, first, to develop a variety of treatment approaches, prescribing as well as non-prescribing, and second, to enhance services outside London.[49]

## 4.3   Doing something about the problem:
## The Central Funding Initiative at work

Central funding for voluntary organisations working in the health field, as discussed in earlier chapters, was not in itself a new development. In 1982–83, (the year the CFI was introduced) the DHSS provided a total of £15 million to voluntary groups operating around almost every conceivable disease and condition.[50] There was also specific support for voluntary organisations in the drugs field, and the amount made available by the DHSS to these groups had steadily risen. In 1980–81 the DHSS gave drug services working in the statutory and voluntary sector a total of £285,690; in 1981–82, £347,360 and in 1982–83, £433,174.[51] This compared quite favourably with support offered to statutory and voluntary services in the alcohol field. While the total sums of money made available were greater, alcohol services were dealing with a much larger population and their budgets actually declined over the same period: in 1980–81 alcohol services received £1,245,018, but by 1982–83 only £846,420 was made available.[52] The amount of money on offer through the Central Funding Initiative to voluntary and statutory drug services, initially set at £2 million a year for three years, was thus a significant sum.

Yet, it was not just the large amount of cash available that marked the CFI out from earlier DHSS funding: the CFI worked in a different way to previous DHSS funding programmes. The scheme operated by setting aside 'earmarked' funds exclusively for drug services which local authorities and voluntary groups could bid for by applying to the DHSS for a grant. Most funding from central to local government was not usually directed towards a specific aim, but, according to Dorothy Black, it was felt that the only way local authorities could be persuaded to do anything about drug use was to provide funding.[53] As Susanne MacGregor and Betsy Ettorre noted, 'Health Authorities find it difficult to put services for drug misusers ahead of those for, say, kidney transplants or old people.'[54] What was also significant about the CFI was the way funds were issued. Grants were made on a pump-priming basis: agencies in receipt of a grant were expected to find alternative sources of funding once this came to an end. For voluntary agencies, this meant securing the support of the relevant local authority at the time of application.[55] This was designed to ensure that when the fixed period of central funding finished, services would not disappear.[56]

The criteria on which the outcome of applications was decided placed local support at the top of the list, but there was also a desire to ensure that services were spread geographically, spread over the different types of service and distributed evenly between the different types of organisation applying – that is, between statutory and voluntary organisations. The DHSS were also keen that services would meet a need 'as yet unmet' and not be 'counter-productive' by 'glamorising' drug taking.[57] Those involved in the initiative remembered the intensive way in which they operated. A senior civil servant recalled that 'I used to be there until 8, 9 o'clock reading through these things [applications]'. All the proposals were 'read by all the professionals in the drug policy group, which were myself, my nursing colleague, and by one of the administrative officers, and we then came to a consensus about what we could fund.'[58] Such scrutiny was necessary because the CFI was heavily oversubscribed. In November 1983, Fowler noted in a letter to Leon Brittan (the Home Secretary) that over a hundred applications had been received, 'amounting to more than twice the funds available.'[59] Indeed, the influx of promising applications was one of the reasons why the scheme was expanded beyond the initial three-year period, finally ending in 1989.[60]

When it was introduced the CFI had four stated objectives: first, to provide regional and local assessments of the drug problem; second, to improve awareness of the problems related to drug 'misuse' and the ability of people working in this area to help; third, to improve links between health services provision and community provision and finally, to improve the effectiveness of services and their value for money.[61] It was clear from the outset that the CFI was designed to improve services for drug users throughout the nation. The Under-secretary of State for Health and Social Security, John Patten, told a committee of MPs examining the nature of drug services in 1985 that 'the concentration of facilities for drug abuse has been in London and the South East, but it is a growing problem in other parts of the country and it would be wrong if we did not spread these resources around the country.'[62] The pattern of grants made supported this. A total of 188 grants were made under the CFI between 1983 and 1989, and as Patten remarked 'places north of London will not have cause to complain' about the allocation of money.[63] Indeed, Chris Smith, MP for Islington and Finsbury, asserted that CFI funds had 'largely bypassed London', an allegation quickly quashed by the government and not born out by the actual allocation of resources. All of the 14 Regional Health Authorities received some funds, although not equally. Areas thought to have the most extensive drug use (the Thames region and Merseyside) received the most money. The team of researchers employed to assess the effectiveness of the CFI concluded that, 'The allocation exercise seems to have been successful in targeting areas of need.'[64] London did not necessarily lose out in terms of the funding allocation (45 per cent of grants were awarded to projects in the Thames region) rather it was a case of the regions gaining funding for service provision where there had been none before.

Yet the CFI did, to some extent, also mark the lessening of the capital's influence on the general shape and direction of drug services. Prior to the CFI, drug policy and provision had been dominated by the Drug Dependence Units: of the 22 DDUs nationwide, 16 were located in the Thames region, most of these in the London teaching hospitals.[65] As these were run by specialist psychiatrists the DDUs could be seen as being representative of a medical approach to drug use.[66] But, as the number of drug users increased during the 1980s and the effects of drug use were felt within society on a larger scale, questions were raised about medicine's dominant role in drug policy.[67] This was exemplified by a change in the terms utilised to describe drug users from 'addicts' in the 1960s to 'problem drug takers' in the 1980s.[68] 'Problem drug takers' appeared to require the help and support of a much wider range of agencies than 'addicts'. A senior civil servant remarked that the DHSS view at this time was that DDUs should give 'further thought to the sort of treatment programme they should approach; that it shouldn't just be an entirely medical, clinical one.'[69] This resulted in an increasingly multi-disciplinary approach to drug use, in contrast to the medically orientated London-based DDUs.[70]

The type of services supported by the CFI both exemplified and accelerated these trends. Almost half (46 per cent) of CFI funds went to community-based walk in centres; 20 per cent to multi-disciplinary community drug teams; 16 per cent to residential rehabilitation facilities, leaving just 18 per cent of funds for DDUs and hospital based services.[71] After the CFI, drug services, according to the team of researchers tasked with assessing its impact, could best be described as 'pluralistic'.[72] An important aspect of this pluralism was the role played by voluntary organisations. Despite statutory support, most agencies working in the drugs field during the late 1970s and early 1980s were chronically under-funded. The ACMD noted that, 'The non-statutory agencies involved in treatment and rehabilitation rely on an insecure combination of local and central government funding and exist under the constant threat of financial collapse.'[73] But, providing central funds for voluntary organisations in order to prevent them from disappearing was not the sole reason for opening up the Central Funding Initiative to non-statutory groups. A DHSS circular informing regional authorities of the introduction of the CFI stated that its purpose was

> not to remove from statutory authorities the responsibility for providing services and training but, by making additional funds available to them and to voluntary organisations, to remedy more rapidly

than would otherwise have been possible, the inadequacy of the network of services for people with drug related problems.[74]

Fostering the participation of voluntary organisations was vital because,
as Patten told MPs, there was a realisation that 'the problem is not necessarily going to be ameliorated and controlled...by action within the
National Health Service alone.' Moreover, 'A very great deal of expertise,
in terms of prevention and counselling, is in the voluntary sector, not
in the National Health Service.'[75]

Yet, non-statutory groups did not just provide expertise: there was
a feeling amongst DHSS officials that voluntary organisations offered
something statutory authorities could not. A senior civil servant asserted
that voluntary groups 'could be more flexible in what they did' that as
they 'were not tied to a specific service approach...they were more willing to initiate different types of services.'[76] The CFI, by offering substantial funding to voluntary organisations, was designed to make use
of this. Even so, a senior civil servant remarked that 'we were quite
surprised that we got so many applications from the voluntary sector';
clearly developments on the ground had been somewhat invisible at
the central policy level.[77] However, once the DHSS were aware of the
extent of voluntary sector involvement in the field a clear commitment
was made to enhancing its role in drug service provision. This can be
seen in the grants made under the CFI: of the 188 grants issued, 58 per
cent went to statutory organisations and 42 per cent to non-statutory
groups.[78]

Such significant support for voluntary organisations cannot be
explained by necessity alone: this must be related to a much broader
strategy for involving the non-statutory sector in health and social
service provision. The House of Commons Social Services Committee
tasked with investigating drug 'misuse' in 1985 were supportive of
a greater role for voluntary organisations in drug services, but were
concerned about the short-term nature of pump-priming grants, and
recommended that 'adequate funding [be] guaranteed for a number
of years.' Furthermore, although most witnesses to the Social Services
Committee welcomed the CFI, some were concerned that 'it will
produce a pattern of services which is haphazard and not truly reflect-
ive of need.'[79]

This criticism resulted in the DHSS commissioning a team, led by
Susanne MacGregor, to evaluate the CFI.[80] The evaluation of policy
initiatives was becoming increasingly appreciated by the DHSS. Other
dedicated funding schemes run by the Department were also assessed

by professional researchers, with the evaluation usually being commissioned at the beginning of the initiative.[81] A DHSS official noted in a letter to MacGregor that, 'Obviously an evaluation of the initiative as a whole should have been mounted at the beginning; for various reasons it wasn't but the customers feel that there is a useful exercise that can still be done.'[82] While the Department felt that an evaluation would be useful for 'both accountability purposes and for general guidance on the further development of services for service planners and policy makers', it might also be suggested that it was required in order to justify policy change and significant expenditure on a heavily stigmatised group.[83] But, the strategy also marked the rise of the role of evidence in health policy and a developing role for research: it exemplified trends which were to develop later in the 1980s through the directed funding for AIDS and the NHS Research and Development initiative.[84]

The evaluation produced by MacGregor and her colleagues of the drugs CFI highlighted the significant impact that the scheme had on the drugs field. The scheme, they noted, 'was of crucial importance in adding a layer of community services to the previously existing hospital and residential provision.'[85] By 1989 there were 323 dedicated drug services in England, and 229 of these (71 per cent) had been established after 1984 and the introduction of the CFI. Moreover, of the 323 drug services 49 per cent operated in the statutory sector and 49 per cent in the non-statutory sector, with the remaining 2 per cent being private organisations.[86] Thus, by the end of the 1980s, voluntary organisations were just as important as statutory groups in terms of the provision of drug services. This was something long recognised by the voluntary sector itself. Eric Blakebrough, director of the Kaleidoscope Youth and Community Project in Kingston, Surrey argued that, 'Many voluntary agencies in this field have been going longer than the hospital drug clinics. They do not need "pump priming". They have been heaving at the pump for years. They need recognition and secure finance.'[87] While the CFI did not provide secure long-term funding, MacGregor and her team concluded that the initiative had seemed to be 'relatively successful as a pump-priming exercise.' Although there were some concerns about the future funding of voluntary groups supported under the CFI, MacGregor found that by March 1990, 76 of the 100 agencies initially funded by the CFI had secured future funding.[88]

Of course, some agencies lost out just as others gained under the CFI. The DHSS used the introduction of the CFI to shift part of the burden of funding organisations from ongoing grant programmes, such as those provided under Section 64 of the Health Services and

Public Health Act, to the temporary funds offered by the CFI. An official noted that the CFI was 'under attack from projects already provided for in the Section 64 programme. Some of these projects are being transferred to the initiatives as means of achieving cuts in the Section 64 expenditure.'[89] What this meant was that organisations that had been relying on indefinite support from the DHSS were now to receive funds on an unsecured basis. Moreover, groups that had long been in receipt of central funds did not necessarily win additional funding under the CFI. Rowdy Yates, coordinator of the street agency Lifeline, recalled that 'residential units including TCs [therapeutic communities] didn't do too well out of the Central Funding Initiative.'[90] Yates's recollection was backed up by MacGregor's analysis of the CFI allocation: residential rehabilitation services received just 9.4 per cent of funds.[91] Yates felt that the reason for this was because these services were 'so resistant to the new ideas about harm reduction.' Similarly, Roger Howard, former director of education training at the National Association for the Care and Resettlement of Offenders (NACRO) and later director of SCODA, believed that before the CFI 'the money had been going into the residences through the old DHSS monies, so there was limited scope for innovation and development.' In contrast, the new pattern of funding introduced through the CFI, Howard contended, 'was instrumental in the development of the voluntary sector and the development of a whole range of interventions and approaches.' The CFI, he remarked, 'stimulated a whole market experience.'[92]

## 4.4    Making sense of the problem: The CFI and drug services in context

Howard's use of the word 'market' is revealing, as in many ways, the Central Funding Initiative for drug services represents a microcosm of key aspects of Conservative welfare policy in this period. The term 'initiative' was a particular favourite of the Thatcher administration. Numerous 'initiatives' were launched to tackle a range of social issues particularly in the inner-cities. Urban development grants, for example, were designed to foster regeneration by using public funds to pump-prime development in areas such as the London Docklands and Merseyside. Central to these policies was the notion of 'partnership' with private companies and voluntary organisations which would be expected to support projects in the long-term.[93] In the health field, 'initiative' had a particular meaning. From 1982 onwards a number of central funding initiatives were launched in areas where the government wanted to raise standards. Alongside the drug users initiative there were CFIs to provide improved services for children under five-years old, better services for mentally ill elderly people, enhanced services for mentally disabled children and more general initiatives such as Opportunities for Volunteering and Helping the Community to Care.[94] A DHSS official noted that, 'The funding of schemes is deliberately limited in duration to preserve their development and catalyst role. They are not intended as a prolonged substitute for local funding.' Health and local authorities were expected to find the money for continuing schemes from within their regular sources of funding and voluntary bodies were required to carry on raising their own funds.[95]

Such a scheme cast central government in the role of initiator of new services rather than their long-term funder. The central funding initiatives thus encapsulated a key aspect of the Thatcherite policy of 'rolling back the state': reducing direct statutory involvement in welfare provision by changing the function of the state from that of provider, to manager, but through a command and control model.[96] This transition was later confirmed through the NHS Care and Community Act in 1990. The act created an internal market within health and social care by establishing a divide between the 'purchasers' of services and the 'providers' of these. Local authorities, for example, were able to 'purchase' a particular service, such as a needle exchange for intravenous drug users, from a local 'provider.' The 'provider' could be a statutory, voluntary or private organisation; these were expected to 'compete' within the internal market for the custom of the 'purchaser'. Competition, it was argued, would make services more cost-effective and responsive to consumer demand.[97]

The creation of the internal market, it has been suggested, helped to replace 'welfare statism' with 'welfare pluralism' as a range of groups and organisations took on functions previously performed by the state.[98] Within the 'mixed economy of care' particular significance was placed on the part played by voluntary organisations.[99] The voluntary sector was regarded as being more flexible than the statutory sector and, crucially, more able to enhance citizen participation.[100] Reliance upon the state could be further reduced as individuals would be afforded the 'invigorating' experience of self-help and community care.[101] Of course there is a paradox here – as statutory support for voluntary organisations increased, and was formalised during the 1990s with the introduction of contracts between purchasers and providers – elements of what was distinctive about the voluntary, as opposed to the statutory, sector could be seen to have diminished.[102] Susanne MacGregor and Ben Pimlott asserted that some organisations were transformed into '*de facto* agencies of the state, which financed them and indirectly determined their policy.'[103]

There was also an additional contradiction inherent within programmes such as the CFI which attempted to control the development of services centrally, as this seemed to run counter to the supposed decentralising tendencies of the Thatcher administration. This situation was apparent to officials at the DHSS. Higher levels of central funding for drug services would 'conflict with Government policy on non-interference with local decisions on the allocation of resources.'[104] A draft letter from Fowler to Brittan noted that, 'The [ACMD's] recommendations for greater involvement by central government, through an extension of my department's advisory role and through departmental grants also pose problems for us. I cannot go further on the former than I have already done... without running counter to Government policies on local devolution and central manpower restraints.'[105] As Rodney Lowe has remarked in relation to the creation of the internal market more generally, this did not automatically result in a reduction of statutory funding for welfare and cannot have been said to have 'rolled back the state'. Public spending on social policy experienced just two cuts (in 1977–78 and in 1988–89) and only fell as a percentage of GDP in the second half of the 1980s, before rising again in the early 1990s. There was a remarkable 'resilience of the welfare state', albeit in a different, more pluralistic, form. For Lowe, the internal market was Thatcher's major legacy in social policy.[106] A crucial part of that legacy was in creating much closer relationships between the voluntary sector and the state.

## 4.5   Conclusion

The Central Funding Initiative for drug users thus needs to be understood in relation to two crises in the early 1980s, one specifically in drug policy and one more generally in the political perception of the role of state provision in social policy. The expanding drug problem, the inadequacy of existing services (especially in the regions) and a desire to shift these away from a purely medical or psychiatric approach to drug use were clearly the key factors in the establishment of a dedicated funding programme for drug services, but the shape that this took was influenced by broader changes in the nature of the welfare state in Britain.

The most obvious immediate consequence of the CFI was to create a much more diverse and extensive network of services for drug users throughout Britain than had previously existed. Services were increasingly multi-disciplinary rather than just medical in nature, facilitating the development of a broader understanding of drug use and its social implications. The CFI's support for voluntary organisations and community-based services was particularly significant: these services came to play a key role in the response to the next crisis of the 1980s, the advent of HIV/AIDS and its spread among drug users. As Black stated at a conference in 1989, AIDS was a 'golden opportunity to get it right for the first time' and the spread of syringe exchange was an additional stimulus to a national system which bypassed the DDUs. Had the CFI not been introduced, the basis for this extension of services would not have existed.[107] Many voluntary groups were leading proponents of the harm minimisation approach to drug use which achieved official sanction following the ACMD's report on AIDS and drug misuse in 1988.[108] In a context where AIDS was viewed as a greater threat to public health than drug use, the voice of the drug user in determining the nature of services was accorded extra weight.

The CFI was, therefore, undoubtedly a product of a specific crisis – the developing drug problem in the 1980s – but the CFI also exemplified broader changes in health and welfare policy. Coming as it did, just before the establishment of the internal market the CFI pre-figured many of its key aspects: central government as the funder but not the provider of services, increased devolution to local authorities as purchasers of services and the growing use of non-statutory organisations in service provision. Furthermore, in the role it gave to research and evaluation the CFI was also a harbinger of evidence based arguments within policy which were to become important in the 1990s and early twenty first century. The CFI was not just an innovative initiative for drug policy: it was a pilot for policy initiatives much more broadly.

It is no surprise, therefore, that the CFI also exemplified some of the inherent contradictions within health and welfare policy in this period. On the one hand, the CFI's encouragement of voluntary organisations, and the devolution of responsibility for the continued success of these groups to local authorities, appeared to be a reduction of the role played by central government in the organisation of drug services. Yet, on the other hand, funding for these services was decided upon centrally, maintaining a crucial role for the DHSS in determining the general shape and direction of drug policy. This is analogous to what has been described as the 'command and control' tendencies of the Thatcher administration, where the rhetoric of enhancing freedom and choice actually masked a significant degree of centralisation. As our analysis of the CFI and drug policy in the 1980s has demonstrated, the state was not so much 'rolled back' in this period as 'rolled in': forming a constituent part of an increasingly diverse mix of welfare providers.

This mix was to prove vital in the response to AIDS, as will be seen in Chapter 5.

# 5
# Activism and Health: The Impact of AIDS

The process of rolling back the frontiers of the state which began in 1979 with the election of Margaret Thatcher's Conservative government necessarily involved an emphasis on the voluntary sector. It could fill the spaces that would open up as the state withdrew from providing, but not from funding, services.[1] The Central Funding Initiative and its initiation in the mid-1980s had shown the complexities of 'state funded voluntarism' which such relationships could engender. The quality of such services was to be secured by market mechanisms, extending the use of contracts as substitutes for grants and by substituting consumer pressure for older styles of voluntary action. In the early 1980s there seemed likely to be a new stage in the relationship between voluntarism and the state, one in which the state operated as funding ring master and target setter for a host of voluntary organisations locked into a closer funding relationship with it. Such developments were far from the older ideals of voluntary action and many in the voluntary sector resented them.

Then came HIV/AIDS. This new syndrome had a significance overall for the nature of the voluntary sector, voluntarism and its relationship with the state. This chapter is about the impact it had on the drugs field, which was paradoxical and two fold. First, so far as the voluntary sector/ state relationship was concerned, AIDS appeared initially to reverse the Thatcherite trends. It substituted a period of pure grass roots social action.[2] This then fell prey to the pressures of state funding and incorporation, as AIDS voluntary action grew closer to the state and the state itself set up a new voluntary organisation, the National AIDS Trust, to rival the 'gay' voluntary sector. Whether the state should take this role and whether conversely 'pure' voluntary organisations should take state funds prompted considerable discussion at the time.

Second, AIDS made the drug user more visible and a player in policy. The emergent focus on patient involvement and patient consumerism had an early, often overlooked, manifestation in the drugs field with the 'rise of the user'. Interest in the concept of the 'patient as consumer' had begun in the 1960s, but took on a particular policy imperative during the late 1980s and early 1990s.[3] Following the publication of the White Paper *Working for Patients* in 1989 and the introduction of the 'internal market' within the NHS, health services became increasingly marketised.[4] Within this market citizens were re-conceptualised as 'consumers' or 'service users'. This development had a number of consequences, drawing greater attention to the quality of service provision but also to consumer or user rights – the right to certain kinds of care or treatment, the right to being treated in particular environment by a particular doctor and so on.[5] A key example of this approach was the establishment of the *Citizen's Charter* in 1991, which gave users of public services, including health services, a series of rights and expectations.[6] These were later built on by the Labour government in the *NHS Plan* of 2000, and its desire to create a 'patient-centred' NHS.[7]

The user focused developments in HIV/AIDS, whether through the involvement of gay men in the conduct of treatment and trials, or through harm reduction initiatives, were early indications of this new consumerism. No longer were users or ex-users the hidden players within voluntary services: the changed focus of policy and the emphasis on harm reduction made their views more central, and targeted funding meant that there was money to employ them as 'professional volunteers'. However some long-term players in the drug scene saw this as a false new dawn. They saw drug users as more visible and involved, but also locked into a medicalised and dependent relationship with the state and with services.

How did this come about? AIDS brought a new imperative onto the policy scene which became entwined with drugs, but some drug policy developments also derived from ongoing tensions and relationships within the established field. So there was a double impetus behind the changes in voluntarism and activism. Henceforth, there were in a sense two entwined policy areas – AIDS and drugs. There were also two stages to the activist developments which AIDS stimulated, a first reaction from the mid to late 1980s, and then a further stage with a greater emphasis on a medicalised response from the mid- to late 1990s. Some developments, such as the greater voice of users within services, and decentralisation outside London, were already inherent in the Central Funding Initiative. But AIDS brought a new voluntarism with it, the initial self-help of the gay organisations which adapted to HIV or which were set up specifically in response to the new syndrome. These were run by gay activists, metropolitan middle class gay men, many with a decade at least of campaigning and activist experience. They were a 'target group' very different from the average drug user. But the gay and drugs issues became entwined, in part because of the epidemiology of the syndrome and the importance of drug users as a potential conduit for HIV into the general population, in part because of personal connections between the two worlds. The impact on the drug field was quite profound. The crisis of AIDS brought new organisations into the field, new players both outside and within government; it brought funding and posts in the organisations which built on the earlier CFI funding; and it brought an emphasis on the user and on attracting the user into services. The later stage of activist response in the 1990s placed increased emphasis on user activism and organisation in relation to services and prescribing, and was carried through in alliance with sympathetic medical allies. The resultant focus on prescribing and on harm reduction objectives brought crisis for older models of voluntarism, in particular for the residential rehabilitation services with their ethos of abstention.

## 5.1 AIDS and gay activism

We can look at the nature of the initial gay response and how it gradually entwined itself and impacted on drugs. There were three main areas of interconnection. Gay models of activism impacted on the drugs field; the voluntary / state relationship deepened, even to the extent of the formation of a new national organisation; and the gay involvement in AIDS treatment, trials and research provided a model for the involvement of drug users in drug treatment. In terms of activism, the early gay response to AIDS from 1981 to the mid-1980s was one of pure voluntarism at a time when 'no one knew anything'. Those who took part in the nascent AIDS organisations remember a sense of intense enthusiasm and commitment, a feeling of pure voluntarism and excitement, despite the gloomy and threatening nature of the cause. Janet Green, a volunteer for London Lesbian Line in the 1980s, was a volunteer for the Terrence Higgins Trust and then became in 1985, one of the first two paid members of staff.[8] (The other was

Nick Partridge, subsequently the Chief Executive). The atmosphere was frenetic.

It was chaos – 15 people in one room – all having meetings – phones ringing everywhere ... very exciting, very stimulating, frustrating – it was the best of times, it was the worst of times – I'm glad I was involved at that time.

Green's post title was Director of Clinical Services but she changed it to services co-ordinator; in practice it meant doing everything, writing leaflets, counselling on the phone and in person, supporting and co-ordinating volunteers, cleaning the toilets.[9]

The response built on earlier gay self-help and political activism in the 1970s. A gay self-help group established in the 1970s, the Lesbian and Gay Switchboard, was of central importance for the initial response to HIV/AIDS, organising the first public conference on the subject in the UK in May 1983. It opened a special helpline after a BBC Horizon programme, 'Killer in the Village', was broadcast in April 1983. The Horizon programme and the May conference also led to the refounding of what was to become the key voluntary organisation in the response to HIV/AIDS, the Terrence Higgins Trust. The Trust had originally been established by friends of 'Terry' Higgins who had died of AIDS in 1982. Its first image had been working class, associated with gay bikers and with benefit events in gay pubs, raising funds for research. The image of the Trust changed as a new group of metropolitan gay men took over. Under the Chairmanship of Tony Whitehead, whose history in gay activism went back into the '70s, it developed a middle class image, 'the alternative professionals' as one participant put it, not only with a focus on health education, educating the gay community about the dangers it faced, but also with the aim of influencing government policy.

This period of pure voluntarism also threw up a host of other self-help organisations. Body Positive developed in London in late 1984 and early 1985 out of a Terrence Higgins Trust support group for people who had been diagnosed sero positive through the newly available blood test for antibodies to the virus.[10] There was also a strong self-help response at the local level. A network of gay men developed local responses, often budding off existing gay organisations. Local gay groups in Cardiff, Bristol, Cambridge, Brighton, Oxford, Exeter and elsewhere established helplines, called meetings, tried to obtain funding. In Cardiff, an AIDS helpline was established in 1984, based on telephone counselling which had been run by a gay organisation, Cardiff Friend, since the 1970s.[11]

In Brighton, what later became the Sussex AIDS Centre developed out of a Body Positive group and the local gay helpline. Initially it had no premises but had the use of the local Family Planning Association's telephone and office three nights a week. This pattern of initial self-help building on the gay organisations of the 1970s (telephone helplines in particular) and some loose supportive and funding relationships with health and local authorities (in Cardiff, Oxford and Cambridge, for example) characterised the early voluntarist response. Spencer Hagard, District Medical Officer in Cambridge during those early days and later Chief Executive of the Health Education Authority, reflected in an opening speech to the national conference of AIDS helplines in 1987, about the speed and nature of the early voluntary response.

> The voluntary response to the social crisis presented by the arrival of HIV infection provides a unique object lesson in that the speed with which organisations have been conceived and born, grown and come to maturity clearly reflects the speed with which the crisis has developed... In five short years, voluntary organisations have grown to a position not achieved by others over a much longer period.[12]

This activism also took a particular medical turn which also pre-figured later developments in the drugs field. AIDS activists became involved in medical research through their membership of trial committees. Information dissemination in publications such as the *National AIDS Manual Treatment and Trials Update* which was published by gay men, made them 'volunteer experts' on the progress of science. This involvement was resisted by other groups because of the 'medicalisation' of gay men which it invoked. Oppositional groups resented the supposed incorporation in the world view of 'official science'.[13] Such tensions were replicated later on by arguments supporting and opposing treatment within the drugs field.[14]

The gay response also threw up difficulties as the initial self-help initiative was eroded by organisational growth and the arrival of sustained outside funding. Gay men such as Whitehead who had experienced the activism of the 1970s, recognised that the Trust, along with other organisations, had fallen into a service provision role. Self-help was inevitably eroded by the advent of significant government funding from 1988. This included the establishment in 1987–88 of a government funded national voluntary organisation, the National AIDS Trust (NAT). Initially funded by the Department of Health, its aim, so leading protagonists confirmed, was to take power away from gay organisations such as the Terrence Higgins Trust, whose agenda was seen as 'not helpful'.[15] The recollections of an early staff member with voluntary sector experience vividly conveyed the difference between this model and the earlier self-help initiatives. The staff of the NAT were 'rattling around in huge offices in the Euston Tower. There was nothing to signify that it was the NAT – we were in the Department of Health! ... They knew nothing, and even less about the voluntary sector ... .'[16] This model of 'rolling in to the state' rather than outsourcing state activities was further developed in the drugs field in the early twenty-first century with the formation of the National Treatment Agency as a state replacement for voluntary sector initiatives.[17]

## 5.2   AIDS and drug activism

Gay organisations and the response to HIV came to have a significant impact on voluntarism in the drugs field, but initially the impact was minimal. In 1983, at a Greater London Council sponsored conference at County Hall on what was then called HTLVIII, a workshop on drugs and HTLVIII attracted only three participants.[18] But gay men and those in the drug voluntary sector had personal connections and networks and these gradually drew the two issues together. SCODA was an important initial location for the new responses. Ed Baraccat, a psychiatrist at St Clements Drug Dependence Unit and a member of the Gay Medical Association, worked with Dave Turner, Director of SCODA, to draw attention to the link between AIDS and drug use which was already becoming apparent in the US and the likely implications this connection might have in the UK. Baraccat wrote an initial leaflet which SCODA circulated but there was almost no response.[19] The first full conference on the AIDS and drugs connection was held at the Worlds End health centre in Chelsea in May 1985. By then there was some more interest; many of the London drugs agencies sent representatives. The conference spawned a group which wanted to be involved in spreading information about HIV to the drugs world. One member of this group was Bill Nelles, a gay Canadian drug user, who was also involved with the Terrence Higgins Trust. He had begun a relationship with John Fitzpatrick, a Trust volunteer, and later the Trust's first Chief Executive. Fitzpatrick, like Whitehead and others, had been involved with gay issues since the 1970s. Whitehead had connections with the drugs world through his lover George Cant, a former drug user who had trained as a counsellor.[20] But the Trust at this stage had no drug dimension to its activities. Whitehead and Fitzpatrick suggested that Turner employ Nelles at SCODA in order to produce a booklet on AIDS for drug workers. Nelles began work at SCODA in October 1985 with funding from the DHSS for production of the leaflet. Relationships were not easy despite the existence of a joint SCODA/THT working party. Nelles was taking methadone, which contravened an unofficial rule in the drug voluntary sector that staff should be ex, rather than current users. Nelles remembered that 'THT did not have a problem with this, but SCODA and Dave Turner did. Drug using was not allowed in a drug organisation.'[21]

Nelles epitomised the new tendencies within drug voluntarism and policy which AIDS had engendered. He was a visible drug user. He was also an outspoken advocate of harm reduction, and his speech in February 1986, advocating needle exchange as a harm reduction tactic, was widely reported. This strategy highlighted tensions within the drug voluntary sector. The concept of harm reduction or minimisation had been emergent there for a number of years but caused deep divisions. Those in the residential rehabilitation houses saw their main aim as stopping injecting, not promoting it: their dominant ethos was abstinence. The arrival of HIV among drug users in the UK was revealed in the autumn of 1985 when the blood of drug users in Edinburgh, taken during an earlier outbreak of hepatitis B, was tested.[22] This discovery brought the dilemmas for the voluntary sector response to a crisis point. The confusion of objectives within the traditional drug voluntary sector meant that new structures set up to deal with HIV/AIDS initially took forward the changes which resulted.

One of these was the Drugs Education Group at the Terrence Higgins Trust. Nelles left SCODA in the spring of 1986 and moved to the Trust to set up its drugs work. Relationships were not easy there either. There were tensions between gay men and the drugs group. The Trust was, as one volunteer put it, 'run by gay men for gay men' and so there was resentment at the drugs side. Janet Green remembered that acknowledging that they had to take on support in the Trust for 'those people' caused a real furore. The Trust buddies, volunteers who worked intensively with one person with AIDS offering whatever support they needed, discussed this change. These volunteers stressed that they were there to work with gay men. ' "We don't want to work with rotten old drug users". It took a lot of training and consultation to educate volunteers to see that this was exactly the same discrimination as the gay community were fighting.'[23] Much energy went on internal battles. The

Department of Health knew of poor financial management but felt that the lead agency could not be cut. A leading DH civil servant believed that the Trust was wasting its drugs funding. But the DH AIDS Unit looked to the Trust for its drugs thinking – against the advice of Dorothy Black who had much longer experience within the drugs sector. Further complications came with the move to THT of a drug support group from St Marys, which included a drug user, John Mordaunt. Mordaunt and Nelles clashed over who would run the drug group; and eventually both Nelles and Fitzpatrick left the Trust, leaving Steve Cranfield to run the Drug Education Group.[24]

Subsequently the job of co-ordinating the THT drugs work was taken on by Betsy Ettorre, a sociologist who had worked in the drugs field at the Addiction Research Unit (later the National Addiction Centre). Ettorre too, after an initial good start, found that relationships were difficult and resigned after only seven months in post at the end of 1989. Her departure was followed shortly afterwards by the resignation of the whole drug advisory group, which had, by then, recruited some well-known names from the drug service world. Drug users who came into contact with the Trust's drug work felt similarly at odds with the organisation. A user activist recalled an additional anti-female bias.

> ...even within Body Positive, that's right, cos we went to the Terence Higgins Trust, and we went to the Lighthouse, and even in those places, we started, we really felt, and I really felt, and of course mum felt it with me, just being blanked. Being ignored. And at first I kept thinking, we kept focusing on the issue about it being women, that it's a women's thing, that we're going to these HIV communities that are very gay centred and they're not, they're not accepting of women.... I sort of discovered this at Terrence Higgins, and the Lighthouse and Body Positive this kind of, just a real suspicion and disinterest. I mean they knew you were a drug user, they knew you're a drug user and a woman and the two things didn't fit in with this little gay, tight knit community that they'd set up. And that was a real shock.[25]

The impact of the new AIDS organizations in the drugs world was thus complicated by the structural and personal tensions between the activist traditions of gay men and the emergent visibility of drug users, a heady mix which also caused tensions with the pre- existing 'policy community' round drugs.

## 5.3 Policy change and the user in the 1980s

Changes in the voluntary sector paralleled those wrought in policy by HIV/AIDS. There the impact of the syndrome brought to the fore an emphasis on what came to be known as harm reduction. This was first enunciated officially in Scotland. The Scottish McClelland committee, which examined the situation in Edinburgh and elsewhere, reported in September 1986. It gave a clear enunciation of the emergent new orthodoxy in drug services and treatment. This was that reducing harm was more important than eliminating drug use. It was important to keep drug users in contact with services to reduce the danger of the spread of HIV. The report stated

> There is...a serious risk that infected drug misusers will spread HIV beyond the presently recognised high risk groups and into the sexually active general population. Very extensive spread by heterosexual contact has already occurred in a number of African countries...There is...an urgent need to contain the spread of HIV infection among drug misusers not only to limit the harm caused to drug misusers themselves but also to protect the health of the general public. The gravity of the problem is such that on balance the containment of the spread of the virus is a higher priority in a management than the spread of drug misuse.[26]

This statement was echoed later by the Part 1 report of the Advisory Council on the Misuse of Drugs report on AIDS and Drug Misuse which was published in 1988. The report used the phrase 'a hierarchy of goals' to frame the appropriate response to the treatment of drug users in the wake of HIV/AIDS.[27] No longer was abstinence the key; what was more appropriate were a range of harm minimisation and community-based strategies. These included needle exchange and prescribing as a bait to attracting drug users into services. The implications for services and for the view of 'the user' were profound and a significant change from what had gone before. For in the 1970s drug treatment had moved towards a 'one size fits all' abstinence-based model, which users could take or leave. This approach had replaced an emphasis on prescribing after the 1971 Misuse of Drugs Act had operationalised the conclusions of the second Brain report. Methadone had replaced heroin in the 1970s, but was usually available as part of a short-term reducing contract of treatment focussed on abstinence within a defined period of time.[28] AIDS brought a new dimension into this side of drug policy. The user had to be attracted into services, possibly with the bait of prescribing. Services had to become more user-friendly. Users began to become more visible at the level of policy and practice.

This was a national policy but the initiative was also matched at the local level. A particular catalyst was in Liverpool. The city was a volatile mix in the 1980s with the left wing Militant in power in the local council; a drug consultant, Dr John Marks, in the Drug Dependence Unit, who publicly opposed prohibition policies and championed an early version of harm reduction; a 'drug epidemic' on Merseyside which attracted national attention and a long-running programme of research, together with a strong connection with emergent 'new public health' thinking at local and international levels.[29] Prescribing and needle exchange were both pioneered at this local level, and drug users who were open about their former use were centrally involved. One key figure was Allan Parry, an ex-user who worked with Howard Seymour of the Regional Health Authority to establish the first needle exchange. Parry became a public figure, speaking, alongside others from the Mersey experiment, at a range of conferences and meetings. A user group, the Mersey users group, was established in connection with John Marks' clinic.

The Liverpool experiment was feeding in through media publicity and publication and there were user groups establishing themselves elsewhere as well. One of the earliest drug user groups was Nottingham Drug Dependents Anonymous (NDDA), which formed in the mid-1980s.[30] NDDA was 'established without encouragement from professionals, the impetus coming entirely from the commitment and determination of the drug users who set it up.' It provided support and guidance for drug users and their families in the Nottingham area and also provided a 'strong "clients" voice for the kinds of services which they themselves want and believe ought to exist.'[31] A long-standing user activist remembered this early group in Nottingham

A guy called Dave Cameron. Who I believe is dead now. But Dave Cameron was a big SCODA member and attended a lot of those early SCODA groups...yeah, Dave was kind of quite helpful. And I thought a very mature kind of user organisation and a very kind of mature vision of what a user organisation should look like, you know. And so you know, he was registered as a charity, he didn't like, a board that was kind of made up from professionals, carers and users, you know.[32]

In Manchester, MAINLINE (the Manchester Addicts Information Network) also looked to a patients' union model. In London at around the same time, the Drug Dependency Improvement Group (DIG) was created. The group was set up by drug-using patients of the London-based private practitioner Dr Ann Dally, who faced being struck off the Medical Register by the General Medical Council because of what was seen as her unorthodox treatment of heroin addicts. DIG aimed 'to lobby to defend Dr Dally in her appeal' but also 'to seek a replacement source for long term addicts to obtain prescriptions/support etc.'[33] DIG thought of themselves as a ' "patients' union" to speak on behalf of drug users.'[34]

In London, the mix of AIDS and drugs led to a drug specific AIDS activist self-help organization – Mainliners, which was established in Brixton with funding from the National AIDS Trust.[35] A user activist who helped set up the original office described the excitement of getting 'official' funding:

Ian was the founder of Mainliners. Yeah. Gay man, ex-user. Most of the group were HIV positive as far as I recall, yeah, yeah. Anyway, so then Ian made this application to the National Aids Trust and got £10,000. It was like, wooh, they gave us £10,000, you know, it was amazing. So we went and set up an office in Brixton Enterprise Centre.[36]

Mainliners eventually moved into the Trust and the drug user group were an important conduit for their type of knowledge in both the developing AIDS sector and in the established drugs field. Drugs workers knew little about HIV and people in the AIDS field needed to be educated about drugs:

[P]articular things we knew that the HIV field didn't have a clue about. You know, for example, about the law. And so for example John and me did a training for the Buddies of the Terence Higgins Trust and they were very eager to help buddying drug users, but because drug users often had illegal drugs in their house that sort of put them off, because if there was a raid they might be implicated. But some Buddies did take it on, and the worked out their own rules. Because you know at the end of the day if they happen to be there in a raid, they're not, you know most of the time the police aren't really interested in that anyway they're just interested in who's the dealer and so on.[37]

The users had special knowledge which both fields lacked at this time and this gave them a status of 'lay expertise' which was something quite new for a drug user. Hearing the 'voice of the user' would have been unthinkable at a drug conference prior to the advent of AIDS. But in the late 1980s and early 1990s conferences began to develop their regular 'user slot' and the experience from the grass roots was accorded a recognition which it had not had before. At a conference on drugs and HIV at the then Hatfield Polytechnic in the early 1990s a drug user spoke in halting terms, while, in the audience, Dorothy Black from the Department of Health spoke informally with other users.[38] Users themselves remembered the importance of speaking at conferences as a means of dissemination: the personal testimony assumed a status it had not had before.

This was an emergent activism which was formed through international links and example. The Netherlands with their Junkiebonden (users' unions) had long provided a type of model which some commentators thought could be imported into the British situation. Groups cited the Dutch Junkiebonds, or drug users' unions, as important sources of inspiration.[39] During the early 1980s the Junkiebonds appeared in a number of cities across the Netherlands, offering harm reduction initiatives such as needle exchange alongside lobbying on treatment issues.[40] Drug user groups also began to appear in a number of other countries during the late 1980s and early 1990s, in countries such as Germany, (Deutsches AIDS Hilfe and JES or Junkies, Ex-Junkies and Substitute Users) Denmark (BrugerForeningen) and Australia.[41] Much of this activity was inspired by HIV/AIDS and the need to introduce harm reduction measures, but many of these international groups continued to exist and exert influence. The Dutch psychologist Ernest Buning was an important connection in the early days of AIDS, showing Norman Fowler round Amsterdam, and also pulling together a European network and newsletter about methadone prescribing – *Methworks*.

The influence which users remembered also came from a more traditional drug connection – with the US. The American sociologist and activist Sam Friedman had written about Dutch user groups as part of a mission to try to develop similar user organizations in the US. His writings were widely disseminated and influential in both Britain and the US.[42] Mainliners at the Trust also developed connections with other European user organizations. One user activist remembered connections with a German user organization which led to funding for a short-lived European interest group of drug users.

The other thing that we did a lot of every time we met, we met about 4 or 5 times a year in different European cities, we used to have press conferences. They were fun. Werner was a great orator, really banging the drum about getting this issue out in the world and you know, educating the press who were also just getting on to it I suppose. Politicising people, actually. Cos I think that was the main thing. I think Werner was very important around...actually the European Interest Group of Drug Users who were the first people that I'm aware of anyhow, to ask for a renegotiation of all the drug conventions.[43]

These new European and international networks paralleled those being developed in the AIDS field more generally: and the influence of American AIDS activists on the UK gay AIDS response had been profound.[44] AIDS had ushered in and stimulated a new era of European and international activism and user involvement. In the absence of later technological developments such as the internet, the conference was an important mechanism of knowledge dissemination and consolidation.

## 5.4 Policy change and the user in the 1980s and 1990s: Professional voluntarism and the rise of treatment

The early response had been one of pure voluntarism but the relationship with the state and with government funding became increasingly important and inevitably changed the nature of the initial activist response. These new self-helping activist organisations were usually supported by a degree of the state funding which became available by the end of the 1980s. The voluntary sector response settled in to a recognisable mixture of voluntary sector /state funding. A raft of new organisations and posts were funded. The AIDS funding came just in time and built on the earlier Central Funding Initiative money. A host of different organisations operated at the local level but these were mostly funded by government grants. New posts were also established within services. Outreach workers were appointed under HIV funding and HIV prevention policy co-ordinators at the local level. In 1989, District Health Authorities were instructed by the DHSS to appoint these co-ordinators, focussing on joint health authority, local authority and voluntary sector strategies for HIV prevention. The proliferation of such posts could be confusing. In Haringey, for example, the local HIV co-ordinator working in the local authority HIV/AIDS Unit liaised with a health prevention team led by a separate District Health Prevention co-ordinator. A separate drugs advisory service was led by a drugs co-ordinator. The proliferation of these posts at the local level foretold the later drug specific initiatives at the local level and the formation of the Drug Action Teams, established in 1995. The new workers were often people who had moved over from the voluntary sector or from volunteering and who became 'professional volunteers'. AIDS in general engendered an ethos of activism, both for gay men and for drugs.

The arrival of 'professional volunteers' symbolised a change in ethos. There were other changes which brought greater involvement for both gay men and for drug users in a more medical role. Increased user involvement in research and in service provision marked a new stage in the impact of HIV/AIDS. For gay men, this brought the development of user-friendly services and the modification of the randomised controlled trial to take account of user views and the speedy application of positive research results. Treatment activism on the US model was more limited in the UK and ACTUP's activities were more circumscribed, after a well-publicised demonstration at the Wellcome AGM in 1990. Mainstream AIDS activism among gay men in fact sought not to challenge 'official science' but rather to seek to improve its functioning and to make the results of research more quickly available.[45] For drug user activists, access to clean needles or to methadone became the primary aims. Drug users, like gay men, were also increasingly involved in research, but their involvement was as researchers themselves, helping research projects to gain access to this 'hard to reach' population. The new Centre for Research on Drugs and Health Behaviour funded by HIV/AIDS money and headed by Gerry Stimson was particularly known for its use of user researchers. Activism coalesced round a medical response to AIDS which brought drug users and their doctors together in an alliance. This alliance indeed had echoes of earlier times. In the 1920s the Rolleston committee on morphine and heroin addiction had, in its report, cemented a pre-existing alliance between primarily middle class addicts and their doctors and legitimised the prescription of opiate drugs to them and this doctor/patient relationship had in fact characterised British drug policy into the 1960s.[46] In the 1990s users were by no means primarily middle class but the focus on prescribing and a drug centred solution again came to the fore.

In the 1990s, tensions within the drug field came to a head over this issue and further policy reports advanced the medical case. The residential rehabilitation organisations, with their ethos of abstention, were severely challenged by the advent of HIV and the resultant emphasis on prescribing. These organisations also fell foul of provisions in the NHS and Community Care Act of 1990 which were implemented in 1993. A worker at SCODA, where these organisations played a significant role, described how special earmarked funds had been intended for them within the original Bill. The worker recalled,

And when the Bill came out, it was revised, and we had the ear-marking in. Which protected residential services. And then for some reason, we've never known why, this was changed, this was suddenly dropped. Maybe they didn't like it, no government ever liked the idea of earmarking funds anyhow. And the Local Authority Associations were upset about it, we were upset about it, obviously. And we had another meeting when it was all the representatives from the local authority associations, and from the voluntary sector with Virginia Bottomley, who was Secretary of State, Michael Portillo, who was Minister of State at the time, and Mahwinney, Brian Mahwinney. And I'd already had a private meet-ing with Brian Mahwinney arguing the case. We argued again, and didn't succeed.[47]

SCODA's co-ordinator, David Turner, went public and fought a strong campaign, in the process undermining European Drug Prevention week in 1993, which was dominated by this issue. It was this battle which some saw as having undermined Turner's position in the drug field and left him open to marginalization later on, as we discuss in Chapter 6. The episode also served to undermine SCODA's position as a body representing the drug voluntary sector and hastened Department of Health determination to 'cross the divide', to start to bring the co-ordination of voluntary and statutory services together.[48] Older styles of voluntary activity in the drug field were less significant and a new era had begun. Government involvement in the voluntary sector became more direct.

The community care debacle was also part of an overall question-ing of treatment and its effectiveness by the Minister of Health, Brian Mawhinney, a medical physicist and Belfast Protestant, who did not favour the new harm reduction ethos. A Department of Health civil servant, speaking in 1993, commented: 'He's trying single-handedly to reverse the policies of harm minimization towards abstinence and moral counselling. We all keep hoping that he'll be reshuffled and then we can sit back and go along as before ...'[49] He instituted a treatment effec-tiveness review which reported in 1996 and which was accompanied by the results of a research study which had an important impact on the field. NTORS, the National Treatment Outcome Research Study was directed by Michael Gossop, a long-standing researcher in the drugs field, and its results could be summarized in two words – 'Treatment works'.[50] The Mawhinney initiative thus had almost exactly the oppo- site effect to its initial intention. The primacy of a medical approach was reasserted within drug policy, but the importance of the connec- tion with a criminal justice agenda was also underlined and was an important part of the new policy package. Treatment was an important diversion route for drug users out of the criminal justice system and a way of potentially breaking the drugs/crime link. It was also something on which (nearly) all could agree. The treatment effectiveness review chaired by Dr John Polkinghorne, a physicist and priest who was head of a Cambridge college, emphasized the need to attract users into serv- ices and reaffirmed the earlier ACMD report's emphasis on the need to prescribe.

A second stage of user activism began in the late 1990s. Polkinghorne had emphasized the need for GPs to become involved in treating drug users. This became a feature of the new era; again, there were already local initiatives on the ground. One was in Brent, where Dr Chris Ford had been treating drug users since the 1980s. After she began to pre- scribe for a user who came to her from St Marys DDU, she was soon approached by others. In fact her practice was said to be 'like a drug service with a general practice attached.'[51] One drug user who came early on was Beryl Poole, who had been treated at a DDU but wanted to change. About five drug users got together and decided not to abuse Ford, their new doctor. They advised Chris Ford to insist that they picked up their drugs on a daily basis. 'If I say the dog has eaten it, you don't have to believe me.' Out of this initiative developed an early grass roots drug users group, the Brent service users' forum, which was men- tioned as an influence by many user activists in the 1990s. There was an equalization of the balance of power between doctor and patient. 'It set off a dynamic that we were equal.' Ford's initiative helped to stimulate a rise of general practitioner interest in drug users (also underpinned by the new clinical guidelines). The Royal College of General Practitioners began to hold regular conferences for GPs on the subject from the mid- 1990s: the substance misuse group in general practice was established in 1995.[52] There was tremendous interest and Beryl got to speak in pub- lic for the first time – a significant rite of passage. Someone from the DH had been on the programme and was 'absolutely appalling'. Beryl got angrier and angrier – 'the last speaker has really pissed me off' – and the GP audience were on her side.[53] The conference always had users and user sessions.

On to the scene in the late 1990s came a user organization called the Alliance, whose Trustees included Chris Ford and Gary Sutton. Originally the Methadone Alliance, on the model of the Dutch network of the 1980s, it dropped the methadone in its title early on to take into account the fact that not all drug using patients were prescribed methadone. With the support of Clare Gerada, the GP adviser in the DH, and funded by section 64 money, it concentrated on advocating for individuals and on working within services. Bill Nelles again emerged as a key figure, once again epitomising some of the new relationships in the drugs field. Nelles had worked in Berkshire within the NHS in the late 1980s on a GP training programme in methadone prescribing. In the 1990s he held posts as HIV co-ordinator in Birmingham and in Harrow. Then he went back into patient involvement work. After speaking at a South Bank University conference, 'narrowing the divide' in 1997 or 1998, he found 'all sorts of activists I remembered from the early days of HIV-patient participation was back [on] the agenda.'[54]

But patient consumerism in the drugs field was problematic. Nelles' pragmatism and ability to compromise and work with different interests was not to the liking of all, nor was the Alliance's work around the supervised consumption of opiate substitute drugs. Most treatment providers argued that supervision was needed in order to prevent overdoses and the 'diversion' of drugs to the black market.[55] But this was inconvenient for users who were forced to attend the service or pharmacist on a daily basis. Some also saw it as demeaning. At a local level, the Alliance had some success in getting service providers to alter their policies on supervised consumption. In the Midlands, it seemed that supervised consumption had been introduced in treatment services without warning or consultation with users. Representatives from the Alliance met with local service providers and an appeal mechanism was devised. As a result, around a half of users moved back onto take home doses. For those that remained on supervised consumption the reasons and alternative were fully explained.[56]

At the same time the Alliance was trying to influence national policy on supervised consumption. One user activist contended that Bill Nelles, the General Secretary of the Alliance, had decided that supervised consumption was a 'lost battle', that 'It's going to happen, accept it and let's fight one we can win.'[57] Nelles himself saw the situation slightly differently. He argued that deaths from methadone overdoses were a problem and that users would need to work with government to put together 'a supervised model that was reasonable and fair.' Nelles asserted that this resulted in a 'very benevolent policy' and that 'all of this has come about because of a much greater collaboration between the users who know how to reach users, and talk to them, and government who understand that we can be worked with, and [that] we can be constructive stakeholders.' This approach, he said, had helped 'the Department of Health to see us as responsible players' and that being 'reasonable' had enabled the Alliance to gain access to the policy-making process. Nelles suggested that, 'Once they [the NTA] were clear that we were not going to subvert government policy and we were focusing on treatment and quality, they were sort of saying this is really quite helpful to us.'[58]

However, the Alliance's work with government on supervised consumption was less popular with some other user activists. Nelles came in for 'a lot of personal criticism at the time...the view was that I was colluding with something that was oppressive to drug users.'[59] For some user activists, collaborating with government over supervised consumption was an unacceptable compromise. One stated that

> the first chance we get to battle we decide that we're going to loose, so we're gonna move on to something else. Eventually we'll end up arguing about our right to, I don't know, not to have to wear a tie to turn up to counselling or something...you'll just be marginalised to meaningless arguments.[60]

Indeed, the Alliance was obviously unable to prevent the introduction of supervised consumption – in 2006, 73 per cent of services supervised users' consumption of drugs in the initial stages of treatment – but this episode does indicate that drug user organisations such as the Alliance were being listened to.[61] They may not have been able to change the overall policy, but they were able to shape its introduction. The Alliance's 'pragmatic' response to supervised consumption also helped the organisation to win the respect of government, enabling the group to gain access to future policy making.

## 5.5   Conclusion

The Alliance's approaches were criticised from a different perspective by long-term workers in the drug field. They saw its focus on improving access to drugs and prescribing as ultimately demeaning and preventing a more active involvement with getting people off drugs. One long-term activist characterized the AIDS period and its aftermath in the following way:

> Theories of addiction were at their lowest and the DDUs stagnated, the residential units didn't do well...they were resistant to ideas about harm reduction...Non medical, non residential self help stuff appeared to offer more. Agencies of that kind got most of the money...But along came AIDS and the medical profession stepped back centre stage. Fewer drug users were more visible, it was a badge of honour, it never was in the 1970s, we never talked about being ex users in our organization. It's proclaimed now, but you rarely see anyone in a position of power, they're all taking rucksacks of needles round housing schemes.[62]

Another saw the new agenda as ultimately destructive for the user.

> The CDTs [Community Drug Teams] are now part of the crime reduction apparatus.... What happens to the person – their alcohol use increases – they put on weight and they get type 2 diabetes. Methadone wine and welfare is now part of the...landscape....[63]

Drug users 'came out' into public visibility for the first time as a result of HIV/AIDS, but a decade later, there were divisions within the drug field about what that enhanced visibility really meant. A debate had begun about the purposes of service provision and by implication of activism which became more public in the early twenty-first century.[64] For both gay activists and for drug workers and users the 'new dawn' of AIDS had brought with it classic dilemmas which affected the relationship between voluntarism and the state. In a speech given at the Institute of Contemporary Arts in 1988, Tony Whitehead of the Terrence Higgins Trust, had articulated the dilemma.

> instead of pulling the rug from under the government, we said, 'Yes, we must do something. We must strengthen the voluntary sector...Our immediate response to the tragedy of AIDS has been to rush off to hold people's hands at bedsides. We have not taken our fight out on to the streets as has happened in the United States.[65]

Gay men had provided services instead of continuing their activist role. For drug users and the wider drug field, the price of prescribing was acceptance of the government's criminal justice focussed agenda.

Treatment was only so widely available and well funded because treating users kept them out of prison, a prime objective of government policy. Treatment 'worked' because it reduced the prison statistics. As we will see in the following chapter, these developments led to new initiatives in the drugs field which renegotiated the relationship between state, voluntary and private sectors. But they also brought a greater role for the state – a 'rolling in to the state' – which was epitomized by the formation of the National Treatment Agency.

# Part III
# 1990s–2000s

# 6
# Business Models or the Revival of the State?

During the 1990s a number of changes took place in the funding and delivery of health and welfare services which had a significant impact on the voluntary sector in general and on voluntary organisations working in the drugs field in particular. According to Jeremy Kendall and Martin Knapp, the Children Act of 1989 and the National Health Service and Community Care Act of 1990 introduced the most sweeping reform of health and social care since the 1940s.[1] The key developments centred on the creation of a market in health and welfare. Local authority 'purchasers' were expected to buy services from independent 'providers': in each region a minimum of 85 per cent of care had to be provided by agencies other than the local authority.[2] This offered voluntary organisations a clear opportunity to compete in the 'mixed economy of care.'[3] Contracts between statutory bodies and voluntary organisations to provide a specific service became the norm, and at the same time, unconditional grant aid was increasingly rare. This can be seen in the overall pattern of statutory support for the voluntary sector. In 1991–92 the VSU gave out £12 million in grants to voluntary organisations, but this figure was dwarfed by the total amount of statutory funding for the voluntary sector of around £11.6 billion, most of which came to voluntary groups through contracts for service provision.[4] The introduction of this 'contract culture' had a significant impact on the way voluntary organisations operated and also on the relationship between the voluntary sector and the state. This chapter will explore the nature of these developments by examining the experiences of voluntary organisations working in the drugs field during the mid-1990s. In the previous chapter we have shown how the advent of HIV/AIDS in a vacuum led to a period of pure voluntarism, a genuine upsurge of activism initially unmediated by state funding or support. This surge of enthusiasm among gay men in turn had its spin off in the drugs world where activism began to take root among drug users. But one of the longer-term results was the expansion of state provided services; many activists secured paid posts and became 'professional volunteers'. In this chapter, we will analyse further developments in these years. On the one hand voluntary organisations themselves changed and began to adopt business models of organisation. This marked a blurring of the boundaries between voluntarism and private enterprise, although still with a strong input from the state with funding through contracts for service provision. On the other hand, the state began to take on activities previously the province of the voluntary sector. These years saw the expansion of the role of the state as an independent operator, in substitution for what had previously been voluntary sector activities. The later history of SCODA, ISDD, the founding of the merged charity DrugScope and the establishment of the National Treatment Agency, which we discuss here, illustrates these developments. State funded voluntarism, the development which has attracted most attention, was not the only change in voluntary sector /state relationships, as these events illustrate.

Indeed, changes in the drugs field during this period tended both to mirror wider shifts in social policy as well as foreshadowing later transitions in other areas. Drug voluntary organisations had already experienced a significant increase in statutory funding before the introduction of the mixed economy of welfare in the 1990s. 'There was a bonanza for agencies in the 1980s' commented one experienced observer of the drug scene.[5] This came in several tranches about which we have written in previous chapters: through the Central Funding Initiative in the wake of the Treatment and Rehabilitation Report of 1982; through the post ACMD HIV funding initiative of the late 1980s; and then in the 1990s came a further influx of investment when the government set up the Central Drugs Coordination Unit (CDCU) in 1994; and also following a review of drug treatment effectiveness in 1996. 'Two themes emerged from all of this' commented another experienced observer, 'drugs became nearly mainstream for health services and there was a plural economy with the voluntary sector as a key provider.'[6] In the late 1980s, the Community Drug Teams and the voluntary sector were jockeying for position, but with the coming of the centrally devised drug strategy and the CDCU in the mid-1990s, it was increasingly clear that central government recognised that what became known as the 'third sector' had a critical role to play in drug service provision.

This shift was occurring at the macro level too. Kendall asserts that in the 1990s the 'third sector' was 'mainstreamed' into public policy, taking an increasingly large share of limited resources and policy-makers' energies.[7] This process was confirmed under the New Labour government from 1997 onwards. The introduction of the *Compact* between the third sector and the state in 1998 highlighted the significance that was now being accorded to the role played by voluntary organisations not only in service provision, but also in the production of a strong 'civil society', a theme that continues to provoke interest and one we will return to in later sections of this book.

## 6.1 Contracts, voluntary organisations and the state, 1990s–2000s

Although the introduction of market-style reform had been a feature of public policy since the mid-1980s, these changes become much more fully realised in the 1990s. The NHS Care and Community Act (1990) effectively created two markets: an internal one within the NHS to provide healthcare, and a somewhat more external market to provide community services such as care for the elderly. Market mechanisms, it was hoped, would improve the cost-effectiveness and efficiency of public services. The spiralling cost of residential care for the elderly in particular drove these reforms forward. From the 1980s onwards the centrally administered social security system had subsidised residential care for the elderly (often provided by voluntary and private agencies) so that by 1992 there were 250,000 claimants which cost the government £2.5 billion.[8] Sir Roy Griffiths, the managing director of Sainsbury's, was called in to assess the situation. His report, *Community Care: An Agenda For Action* (1988), recommended that the responsibility for the provision of care services should be passed on to the local authorities.[9] This proposal initially ran counter to the Thatcher administration's thinking, as policy had been directed towards removing power from local authorities not giving them new responsibilities.[10] However, by 1989 Thatcher changed her mind and the Griffiths proposals were incorporated within the NHS Care and Community Act, albeit with one key change. Following the introduction of the Act, local authorities would not provide services themselves but instead act as an 'enabler', planning the development of services and overseeing needs. The services would then be provided by private and voluntary organisations that would compete for local government contracts. This did not, however, represent the devolution of power from the centre to the periphery: central government would retain a crucial control mechanism by deciding how much local authorities would have to spend on services, 'ring-fencing' funds so that these could not be spent on other things.[11]

The NHS Care and Community Act had significant ramifications for the voluntary sector especially for the use of contracts between local authority 'purchasers' and voluntary sector 'providers' to deliver services. Contract culture formalised the relationship between purchasers and providers and also tended to have a formalising effect on voluntary organisations themselves. Contracts brought with them professional standards of management, assessment and evaluation on voluntary organisations. Lewis found that this caused resentment within some voluntary groups as greater formalisation appeared to undermine what was different about voluntary services.[12] Some regulations imposed on voluntary groups by local authorities might be inappropriate or force the service to alter their objectives, processes or clientele.[13] Other voluntary organisations were concerned that the focus on contracts for service provision would result in a narrowing of their activities, and particularly the demise of campaigning and advocacy work.[14] As Kendall and Knapp observed, contracts made it more difficult for voluntary organisations to remain critical of the government either because of conditions in their contract, or because of self-censorship prompted by political expedience.[15] Indeed, while contracts provided a level of financial security for those organisations lucky enough to be able to secure them, the specific and short-term nature of these created uncertainty and insecurity about longer-term funding.

Moreover, the consequences of contract culture also raised broader questions about the relationship between the voluntary sector and the state. Three rather different approaches to this relationship can be observed in a series of reports on the voluntary sector published throughout the 1990s. The first report emanated from the Home Office and was concerned with the efficiency of the voluntary sector as a service provider.[16] This 'Efficiency Scrutiny', published in 1990, took a very state-orientated view: voluntary organisations had to do what the funder wanted them to do, they had to be cost-effective and they had to do a good job.[17] A later study, also funded by the Home Office but written by Barry Knight for the think-tank CENTRIS, came to rather different conclusions, dividing the voluntary sector into 'first force' organisations which focussed on advocacy and campaigning and 'third force' groups which were effectively publicly funded service delivery organisations. 'Third force' groups, the report contended, had made a 'Faustian pact' with the state and should accept their role as essentially non-profit agencies of government.[18] The CENTRIS report was very controversial, both within government and the voluntary sector, but it did highlight real concern about the impact of contracting on the independence of the voluntary sector.[19]

A more consensual view of the relationship between the voluntary sector and the state can be found in a report prepared for the NCVO by the Commission on the Future of the Voluntary Sector, chaired by Nicholas Deakin. The Deakin report emphasised the independence and diversity of the voluntary sector, and called for the creation of a 'Concordat' between voluntary organisations and the state setting out each party's roles. Deakin seemed to view the relationship between the voluntary sector and the state as a partnership between agencies operating in separate spheres, something critics saw as evidence of an 'earlier backward view' harking back to the Wolfenden report of the 1970s and even to Beveridge in the 1940s.[20] Yet, as Lewis pointed out, in some ways the Deakin report was ahead of its time.[21] A formal agreement between the voluntary sector and the state was later realised with the introduction of the *Compact on Relations Between Government and the Voluntary and Community Sector in England* in 1998, a document which acknowledged the complementary roles performed by each.[22] Furthermore, the Deakin report recognised that voluntary organisations had a wider function than simply either service provision or advocacy; they were also a vital part of a healthy civil society. As Lewis remarked, 'the Deakin report was trying to get across to government that voluntary organisations are not just contractors, but are embedded in civil society with goals of their own.'[23] This was an idea taken up by the New Labour administration which became increasingly interested in the part played by voluntary and community organisations in building a strong civil society.[24]

By the end of the 1990s the relationship between the voluntary sector and the state would seem, therefore, to have been in flux. On the one hand many voluntary organisations were becoming more dependent on the state for funding, and this financial support came with an increasing number of conditions. As a long-standing policy commentator and ISDD staff member, noted, voluntary groups in the 1990s and into the 2000s had 'become more tied into central government.' This, he contended, raised questions about the 'degree to which you can be an independent critique of government... while at the same time being drawn closer and closer together, tied closer and closer in because of funding schemes.'[25] The early 2000s saw a continued growth not only in government spending on the support of voluntary organisations, but also increased recognition of the role these played in service provision. In 2006 a specific government office was set up to deal with the voluntary sector when the Active Communities Unit was merged into the Office of the Third Sector, based in the Cabinet Office. The Office of Third Sector aimed to: 'Develop an environment which enables the third sector to thrive, growing in its contribution to Britain's society, economy and environment.'[26] By the mid-2000s, the amount of statutory funding directed towards the voluntary sector increased to such a level that the government was now the single largest funder of voluntary organisations. In 2003–4 the NCVO estimated that 38 per cent of the voluntary sector's income came from the state, 35 per cent from individuals, 15 per cent from internal sources, 10 per cent from the voluntary sector and 1 per cent from the private sector.[27]

As we discuss in the Conclusion, on the one hand, this growing reliance on the government for financial support has led some commentators to question the extent to which the voluntary sector can remain independent from the state. Yet, on the other hand, there was also a growing recognition of another dimension to voluntary action beyond the narrow view of voluntary organisations solely as service providers. With the revival of debates about the nature of civil society in Britain, voluntary and community organisations were accorded a new significance, one that was reliant not just on their distinctiveness but also on their continued independence.[28]

## 6.2    Drug policy and the drug voluntary sector, 1990s–2000s

The fluctuating nature of voluntary-statutory relations clearly impacted upon the drug voluntary sector, but so did factors specific to that field. Drug policy in the 1990s was characterised by two sets of seemingly contradictory trends. The first trend was an increased level of central co-ordination within drug policy, but this was balanced by developments at the local level such as the consolidation of community-based approaches to drugs and the creation of the Drug Action Teams (DATs). The second trend was the greater emphasis placed on drug related crime and criminal justice measures to deal with drugs. This trend meshed with the enhanced role for drug treatment, particularly around the notion that 'treatment works', as we have discussed in Chapter 5. Henceforward treatment was more obviously tied into the criminal justice agenda. Both these trends were of importance for voluntary organisations working in the drugs field, as can be demonstrated by a brief survey of each set of developments.

Following the dramatic rise in heroin use during the 1980s, central government had begun to take more interest in the drugs issue, as evidenced by the introduction of the CFI in 1982. The 1980s also saw the first attempt to create a coherent response to drugs through the introduction of the first drug strategy, *Tackling Drug Misuse* in 1985. *Tackling Drug Misuse* stated that 'the misuse of drugs is one of the most worrying problems facing our society today' and suggested 'a coherent strategy which attacks drug misuse by simultaneous action on five main fronts.' These were: first, 'reducing supplies from abroad', second, 'tightening controls on drugs produced and prescribed in the UK', third, 'making policing more effective', fourth, 'strengthening deterrence' and finally 'improving prevention, treatment and rehabilitation.'[29] Ten years later, in 1995, a new revised drugs strategy, *Tackling Drugs Together*, was introduced. This document placed equal importance on reducing drug-related crime, limiting the availability of drugs and reducing the health risks associated with drug use. To oversee the implementation of the drugs strategy the Central Drugs Coordination Unit (CDCU) was established within the Privy Council Office.[30] Led by civil servant Sue Street, the CDCU was intended to act as a liaison group between the various government departments with an interest in drug policy and report to the ministerial sub-committee on drug misuse headed by M.P. Tony Newton.[31] Yet, alongside this increased level of central co-ordination some aspects of policy making and practice were also devolved to the periphery through the establishment of local agencies. *Tackling Drugs Together* created Drug Action Teams in each health authority area, designed to tackle drug-related problems at a local level. The DATs were to be made up of representatives from the police, probation and prison services and local authorities.[32]

This trend of developments at both the national and the local level continued under the new Labour government from 1997 onwards. In 1998 a new drug strategy was introduced, *Tackling Drugs to Build a Better Britain: The Government's 10-year Strategy for Tackling Drug Misuse*.[33] This document outlined a number of key changes in drug policy, and has been seen as being representative of a move towards a criminal justice agenda on drugs but *Tackling Drugs to Build a Better Britain* also had important implications for the way drug policy was made and delivered.[34] Under the 1998 drug strategy DATs were given more authority at the local level, and have since taken on the responsibility of commissioning treatment services from NHS and voluntary sector service providers for drug users in their region.[35] Yet, control was not totally devolved to the regions. In 1998 the CDCU was renamed the UK Anti-Drugs Coordination Unit and Keith Hellawell, a former police Chief Constable, was appointed as the UK Anti-Drugs Co-ordinator, popularly known as the 'Drugs Tsar'. Hellawell and his deputy, Mike Trace, were tasked with implementing the drugs strategy and providing 'day-to-day' leadership on the drugs issue.

The appointment of a senior police office as the 'Drugs Tsar' was suggestive of the increased emphasis placed on criminal justice approaches to drugs. Indeed, Karen Duke argues that since the mid-1990s there had been a 'criminalisation' of drug policy, 'a retreat away from the harm-reduction principles of the previous phase of policy...and a move towards the discourses of "crime" "enforcement" and "punishment" and greater involvement of the criminal justice system in drug issues.'[36] For Mike Hough, 'drug-related crime has largely usurped HIV/AIDS as a stimulus in Britain not only for research investment, but also for the expansion of drug services.'[37] Drug users were seen as committing crime to support their habit. In 1995 the National Treatment Outcome Research Study (NTORS) recorded that just over 1,000 drug 'misusers' reported more than 27,000 acquisitive criminal offences in a 90-day period before starting treatment.[38] This finding attracted considerable political and public attention, and has frequently been used to justify the attention placed on drugs and crime. The preface to *Tackling Drugs to Build a Better Britain*, for example, stated that the intention was to 'break once and for all the vicious cycle of drugs and crime which wrecks lives and threatens communities.'[39]

This emphasis on crime led to the police, courts, prisons and probation service becoming much more involved in drug policy and practice than in the past.[40] Commentators such as Gerry Stimson, saw this development as indicative of a 'reorientation of policy away from health and to[wards] drugs and crime.' He argued that a 'healthy' drug policy (in existence from 1987 until 1997) based around tolerance, pragmatism, concern for human rights and 'a consensus between government and those working in the field' had been replaced by an 'unhealthy' policy concentrating on the link between drugs and crime.[41] This case can, however, be overstated. First, criminal justice measures historically had always been part of British drug policy and reviews had pointed out that this emphasis was likely to re-surface at some point.[42] Second, many of the new programmes introduced were actually being directed towards getting drug users into treatment rather than into prison based on the findings of the treatment effectiveness review and the NTORS study, discussed in the preceding chapter.[43] Partly as a result of this finding, a series of initiatives were implemented from the end of the 1990s which bound treatment and crime reduction objectives much more closely together. This could be seen, for example with the introduction of Drug Treatment and Testing Orders (DTTOs) in 1998. A DTTO was a community sentence which compelled an individual drug user to enter treatment as an alternative to other types of sentence, usually prison.[44] Although treatment and criminal justice measures had existed alongside one another within British drug policy since the 1920s, there did seem to have been a greater level of cooperation between the treatment and criminal justice systems from the 1990s onwards. As Duke suggested 'there has been a sea change in terms of the willingness of treatment providers to work within the criminal justice system.'[45]

The increased emphasis on drugs and crime had a number of implications for voluntary groups working in the drugs field, as will be explored in greater detail throughout this chapter. But, it is important to stress that voluntary groups did not necessarily lose out as a result of the growing use of the criminal justice system. Indeed, while the Drugs Tsar was a former police officer his deputy, Mike Trace, had many years of experience of working in the drugs voluntary sector.[46] A more nuanced and complex relationship between drug voluntary organisations and local and national government developed during this period.

## 6.3   The growth of the business voluntary organisation

The large professional voluntary organisation contracting for community-based services became a standard model in the 1990s. Let us look at three: Addaction, Turning Point, and Cranstoun. All had their origin in typical small scale organisations of a recognisable voluntary type, but were transformed in the course of the 1980s and '90s into quite different organisational forms. Addaction, originally the Association of Parents of Addicts (APA) or Association for the Prevention of Addiction, typified the change in the nature and operation of these organizations. It started in the classic 1960s way – through a letter published on the women's page of the *Guardian*. Mollie Craven, the mother of a young addict wrote that 'we parents of addicts are a neglected and ignored group.... I would like to appeal to everyone interested in this agonizing problem to form an association.'[47] The response to the letter was swift and parents' groups were set up under the banner of the APA. In London, Craven and her volunteers ran a rudimentary national office and opened a handful of drop in centres for drug users and their families, funded by grants and private donations. APA helped to spawn other organizations including SCODA and the Surrey based Cranstoun drug services. It focused on training volunteers how to listen non-judgmentally to drug users and gave practical advice on accommodation, healthcare and benefits. When Peter Martin arrived as its chief Executive in 1990, it had four staff, ran two projects in East London and had an annual turnover of £150,000. By 1997 it had became one of the biggest drugs agencies in Britain, and renamed itself Addaction at a glitzy public event.

In part this change was the result of an internal crisis. Earlier in 1997, the *Guardian* had exposed the head of APA's crack project in Nottingham as a drug dealer. The charity denied the allegations and attacked the newspaper. But the case was proved and the man was later jailed for seven years in a separate case of dealing drugs. For Martin, the crisis 'forced us to look at ourselves, to get quality control and to make the organization more accountable.'[48] But much of the organization's growth was also due to the pull of government funding. From 1998, the Labour government's ten-year drug strategy ploughed more than £2.6 billion into drug treatment in the community. By 2007, Addaction was running over 70 projects across Britain, ranging from a mother and toddler service in Scotland to an alcohol and drugs agency counselling young people and their families in Cornwall. Deborah Cameron, its new chief executive, was quoted as saying, 'Around a third of our services are now linked to offenders with drug problems being referred to an Addaction treatment programme as an alternative to prison.'[49]

The expansion of the criminal justice strand in drug policy and treatment also led to the expansion of newer business-style voluntary organizations: Cranstoun, for example, modified the traditional abstention focussed residential rehabilitation approach in favour of a greater flexibility and acceptance of harm reduction. One former service worker remembered an innovative day service called 'The Base' in Wimbledon where the staff comprised nurses, teachers and volunteers who did not insist on people being drug free. The Base mixed users with those who were drug free, which was seen as risky; it had a service user group and offered NVQs, small business advice, a package which would have been highly unusual at the time, but became more common in this new wave of services.[50] Cranstoun, too, funded by the Home Office, began to work with drug users on remand, preparing alternatives to custody to put before the courts. According to her obituary in 2006, Alison Chesney, the chief executive of the organization, had taken hold of 'a macho, unfocussed management team and turned it into a leading provider of harm reduction and treatment services, developing techniques that are now considered routine in drug treatment.'[51] The prison work expanded rapidly in the mid-1990s: the long established Parole Release Scheme was renamed the Prisoners Resource Agency under the aegis of Mike Trace, as its manager, and expanded its work for prisoners either sentenced or on remand. The agency was also concerned with the needs of non-British prisoners in the UK; it had a young offenders team in Feltham prison; and opened a day centre in South West London for young offenders involved in the criminal justice system.[52] One observer talked of a 'quasi privatization of the sector' exemplified with large corporations such as Cranstoun.

This development also marked the transformation of Turning Point and its management.[53] Turning Point began as a small local charity in South London, Helping Hand, founded by Barry Richards in 1964 alongside the opening of the Camberwell Alcohol Project.[54] Further London alcohol projects followed and then in 1969, Helping Hand began to work with London drug users through a project at Suffolk House. In the early 1970s, with three alcohol and two drug projects in the London area, Helping Hand branched out to Manchester. By 1985, it had been renamed Turning Point and was running 16 alcohol and 5 drug services around the country. In that year, it diversified further, with work on mental health services in the East Midlands and in the North West. In the 1990s came expansion into learning disability and also into prison drug services. Its first prison-based service, at Pentonville, was opened in 1997. In 2000 a drug treatment and testing (DTTO) project was set up in Wales, followed by learning disability services. In 2001 Turning Point became the largest provider of the new 'progress2work' employment schemes, opening 12 services country wide. The organization's website proclaimed

> In its 40 years, Turning Point has grown from a pioneering alcohol project in South East London to become the UK's leading social care organization, working in the areas of substance misuse, mental health and learning disability. We have gone from supporting a handful of street drinkers in 1964 to making contact with over 100,000 people in the last year.[55]

Its successive directors and chief executives typified the changes in voluntary sector structures and management. From the late 1970s to 1988 the chief executive was Brian Arbery whose background was typical of health voluntarism in the 1970s. Like Mike Daube, who became director of ASH (Action on Smoking and Health) in the early 1970s, he gained his initial experience in the new single issue politics.[56] Arbery, like Daube, initially worked for Shelter, where the radical activist Des Wilson was training a whole cadre. Arbery then worked in the mid-1970s as the Field Director for the Campaign for Social Democracy, a small organization set up by Dick Taverne, the independent MP for Lincoln, who had left the Labour Party. Arbery's move into Turning Point in 1978 began a period of rapid expansion for the organization, and the development of extensive networks of influence and support among the 'great and good' which became the norm for such voluntary bodies in the 1980s. Diana, Princess of Wales, became the patron of the organization in 1987.[57] Arbery was one of the new breed of 'social entrepreneurs' who married radicalism and a social conscience with a focus on management and business procedures.

Arbery sought to 'tame' the organization, to deal with the resistance of staff to business methods. As one observer put it, drug services were 'the last refuge for the anarchist.'[58] The anarchic workforce was difficult to manage and difficult to professionalize and this led in 1988 to Arbery's eventual departure from Turning Point. Internal management was still an issue when the charity appointed a new chief executive in 2001. Victor Adebowale typified the new style voluntary sector leader. He began in local authority housing when the 'idea that you might borrow tens of millions from the bank was unthinkable.' But the basis of charitable income was changing during the Thatcher era from one based on donations and grants to one based on contract income targets. The marketisation of the social favoured the voluntary sector, but the issue of management had to be dealt with. A senior manager at Turning Point stated that 'I started when collectives were still considered to be a serious structure for the voluntary sector.' The organization was run 'by the staff for the staff.' There were fifty or so people deciding what to do. Everyone would meet every week and the meetings would berate the Director. Turning Point had had four or five Chief Executives in so many years; 'the organization has seen them off.'[59] The collectives were turned into hierarchies.

Turning Point and the other big organizations developed a new 'not-for-profit' ethos, partly because the state was making more money available to service providing voluntary organisations. In 2002, £188 million was allocated to the Active Communities Unit (part of the Home Office) to oversee the development of the voluntary sector and a further £125 million was made available to voluntary organisations through Futurebuilders, a one-off investment scheme for voluntary groups providing services in key areas such as health and social care. As one commentator pointed out, in some areas of need you were now more likely to be looked after by Turning Point than by the local authority. 'The debate about public services – they should be services to the public; there's a blurring of the edges public/ private and not-for-profit- why should there be a difference? The debate should be about quality.' The vast majority of Turning Point's income (99%) came from contracts. Adebowale was an outspoken supporter of partnership working but critical of some aspects of the process. In an interview, he applauded the government for promoting partnership working but was critical of the way in which civil servants had managed the process. 'He calls for "responsible commissioning" and management of the market.'[60] Adebowale, in his work for the Futurebuilders Fund, a Treasury initiative to increase the capacity of voluntary organizations to deliver public services, called for the utilization of government money like venture capital. 'What I would like to see is the Fund used as a means to generate and keep the pot going while investing in big ideas for delivering social change. The risk, he says, is that the £125 million will disappear as a series of large grants, never to be seen again. You have to make a step change for delivering the service. One organization might get £30million to do something but it is not real investment. It goes to one organization and one alone. My approach would be to regenerate the money flow.'[61]

Turning Point along with Cranstoun, Addaction and Phoenix House (now Phoenix Futures), were the 'Big Four' of business partnerships between the voluntary sector and the state. But the tendency towards business models affected many other organizations in the drugs field. The Teachers Advisory Council on Alcohol and Drug Education (TACADE) for example, which had been started by the temperance supporter Derek Rutherford with Quaker and Methodist Trustees, started to change to a business model in the 1980s. ADFAM, the organization for families and carers of drug users, had an origin like that of Addaction. It was set up in 1984 by the mother of a heroin addict. In the 1990s however, it became a limited company and it began to get criminal justice work as did the other charities. But core funding from the DH went in favour of project funding. A worker commented, 'The relationship with government was always 'iffy'. Now the pressure is encouraging the growth of really big organizations...bids are now a mix of voluntary and commercial business organizations. It's the toughest time I've known in the voluntary sector.'[62]

Social enterprise was another form of alliance between business and voluntarism. Social enterprise models varied but common to all was the idea of 'trading activity with a social purpose-value led and market driven'; this was articulated in 2003 by a report produced by the Social Enterprise coalition with a foreword by Tony Blair.[63] There were longer established traditions of mutualism and cooperation in business going back to the cooperative movement in the nineteenth century. Social enterprise as a 'marketing device' came onto the scene in the early twenty-first century. In the drugs field there had been an early organization of this type, HIT in Liverpool in the 1990s. HIT was established at that time as a 'privatised' North West Drug Training Unit. It was very successful in establishing the international harm reduction conferences and in promoting publications and training.

It was a role model for fully fledged social enterprise in the drug field, exemplified in Exchange Supplies, an organisation which evolved in the early 2000s. Exchange Supplies developed out a 'pure voluntarism' activity which then developed a commercial dimension. In the late 1980s Andrew Preston, a student nurse, developed a guide to safer injecting, *What Works?* which was published and marketed by DrugScope. While still working as a community psychiatric nurse, he wrote a whole series of drugs publications which were marketed in the same way: among them was *The Methadone Handbook,* which became the self-help bible for the field. Exchange Supplies began to develop as a social enterprise in 2001 through the issue of citric acid sachets for needle exchanges to supply to drug users. Preston and his co-worker Jon Derricott, who had also worked for HIT and was a founder member of the International Harm Reduction Association (IHRA), decided to take the initiative.

> We had to develop the citric ourselves because the legal problems (and commercial risk) meant that despite our urging, free advice and encouragement the companies, charities and voluntary bodies serving the harm reduction field were either unable or unwilling to fully respond to the equipment and information needs of drug users. As independent trainers we were able to invest our own time and effort and money in finding solutions to the problems surrounding paraphernalia supply. We had begun to hear more frequent accounts from people on our courses of eyesight problems being caused by the use of lemon juice (rather than a more appropriate acid) and became increasingly determined to do something about it.[64]

Their response built directly on the needs of users, and further initiatives followed: among them were the supply of a water ampoule intended for injecting, and the organization of the annual National

Drug Treatment Conference, notable for its involvement of users in the programme. Andrew Preston recalled: 'it was high risk but there was a demand and we added other products.'[65] He signed up for business classes and later they became a limited company, recruiting an employee who provided proper financial planning. There was no grand plan. The company employed drug users and much of their product testing was with staff. In the 2000s many of the larger drug organizations also began to use the social enterprise terminology – Turning Point for example, began to call itself a social enterprise organization. Here was a further illustration of eroding boundaries between public, business and voluntary sectors.

## 6.4 The role of second tier organisations: The revival of the state?

The new relationships with business and the use of business models were not the only direction for voluntary/ state relationships at this time. The state also began to carve out a new more dominant role for itself especially in the field of national co-ordination and advice on treatment. There, it was intimately tied into the fate and role of 'second tier' organisations, those who provided information and co-ordination. Government was interested in the role of second tier organisations within the voluntary sector. In 2004 a policy initiative called ChangeUp was introduced, which was designed to build capacity in the voluntary sector by providing £80 million to develop its infrastructure. In the drugs field, however, the histories of SCODA and ISDD in this period and the arrival of the National Treatment Agency saw these voluntary sector agencies losing power, influence and state funding in favour of greater state involvement. Government took back into the state activities which once would have been the province of the voluntary sector. The story had some echoes of the earlier 'rolling into the state' which had taken place in the late 1980s at a time when the Department of Health was nervous about the role of the Terrence Higgins Trust and had set up the National AIDS Trust.[66] This time it was SCODA and ISDD which fell victim.

SCODA had struggled with some of the changes of the 1980s and 90s. AIDS and the advent of harm reduction as a strategy had challenged the ethos of abstention central to the residential agencies which were its bedrock. The new voluntary organizations, like needle exchanges, challenged the philosophy of abstention. The boundaries between voluntary and statutory services were also becoming blurred: the establishment of Community Drug Teams as part of the NHS seemed to roll the voluntary localised model into NHS services. By 1991, CDTs had been established in more than half the 192 District Health Authorities.[67] Developments at local level with the formation of the DATS and DRGs (Drug Action Teams and Drug Reference Groups) in 1995 spoke of a new alliance between statutory and voluntary sector organizations. The boundaries between voluntary and statutory provision were increasingly blurred.

In 1988, SCODA's management committee had opposed a move to individual membership open equally to the statutory and voluntary sectors, the structure adopted by the Scottish Drugs Forum. But its 1994 AGM voted to open membership to all 'not-for- profit' drug service providers, ending the rule which had allowed only organizations from the voluntary sector to be members, and individuals from the statutory or private sectors. The magazine *Druglink* commented, 'The change promises to give England and Wales a more broadly representative voice for the drugs field as the pressure on services mounts after the halcyon days of the '80s Central Funding Initiative.'[68] The argument was that the advent of 'purchaser/ provider' services meant that they were all providers now. But hardly had this motion been passed, than the residential services mounted a powerful counter attack, arguing that this policy change should not be implemented until the position of the residential services within SCODA had been resolved. The organisation's management committee, headed by Jane Goodsir, a former director of Release, had achieved a policy change desired by the Department of Health, SCODA's main funder, but was unable to implement this until the residential service issue was dealt with.

The Department's concern over SCODA and its ability to represent the field dated back at least to a DH review in 1991. A memorandum from the then Department of Health official Margaret Jackman asserted that

> There have been concerns about SCODA's ability effectively to fulfil the role of a national voluntary body in the drugs field for some years. The Department commissioned a review of SCODA in 1991, wide-ranging recommendations were made, but there has been little evidence of the organisation effecting the required changes.[69]

DH medical advisor, Dr Michael Farrell, noted in 1992 that 'it would be helpful for SCODA to clearly represent the whole of the voluntary sector in the drug field and to have a mechanism by which it clearly achieves a wide representation of views.' However, Farrell did concede that 'given the disparate nature of the voluntary sector this is no small task.'[70]

SCODA's lobbying against the community care provisions had won the organization no friends within government, and matters came to a head in early 1994. The Department made a 'hands on consultancy' a condition of the renewal of its grant to SCODA for 1994–95 and withheld £40,000 to pay for this. As before the department's concerns were over 'lack of credibility' in the field and SCODA's 'limited membership', meaning it was not seen as effectively representing drug services or their clients. A draft document specifying the need for the employment of a consultant to work with SCODA contended that 'there was evidence of SCODA's increasing lack of credibility among organisations in the field, which has adversely affected its ability to increase its membership. There is a perception that SCODA has lost its way and is not having the impact on, or providing effective representation for, its drug service constituency.'[71] After the SCODA AGM, this consultancy was imposed but was not unacceptable to the SCODA management committee which had already been looking to the new criminal justice agenda as a potential focus for the agency's work. Dave Turner resigned as Director of SCODA when his management committee decided to support the consultancy.

An anonymous DH spokesperson speaking to Mike Ashton of *Druglink* in early 1995 said that the change was necessary to align the vision of the organization with that of its major funder. 'If services as in *Tackling Drugs Together* are expected to participate in a wider community based response to drug misuse...it is right that their representative body should reflect that shift.' As Ashton pointed out, SCODA members were concerned to preserve the ethos of voluntarism. 'At stake for many is not just a national lobby for their voluntary sector services, but the closest thing Britain has to a national voice for problem drug users, mirroring its members' traditional client advocacy role'.[72]

Those credentials as an independent advocacy organization took a sharp knock in November 1994, when the external consultant appointed to review the organization, Roger Howard, was appointed as its new chief executive. The appointment caused a furore and some saw it as a new form of control role by the Department of Health. But as Ashton perceptively observed,

> concluding that SCODA was forced to toe the official line would be too simplistic. The reality may be closer to a complex alliance between the Government in the form of the Central Drugs Coordination Unit

(CDCU) and the DoH, 'modernisers' on SCODA's committee, figures in the drugs field, some themselves on SCODA's committee, and the consultant, author of the *Across the Divide* report which informed the CDCU's work.[73]

Howard's report encompassed in its title the new relationship between voluntary and statutory services.[74]

Promised opposition to the changes, which included the redundancy of all core staff, evaporated at the 1995 AGM when concerns about the nature of the organization and its role were deflected into a constitutional review group. The Executive Committee's proposal for the group to report within a year on membership, governance and consultative matters was accepted, but there was ongoing concern about SCODA's stance and its role. Alison Chesney of Cranstoun drew attention to the fact that a *Statement of Purpose* which had emerged from extensive consultation 'makes no mention of serving the needs of drug users, protecting their interests or upholding their civil rights – an amazing omission.' Roger Howard's response was that SCODA was 'not about directly serving the needs of drug users. The field has emphatically said that that is the business of drug services.'[75]

But in practice, SCODA did become involved with the issue of user rights in the wake of the publication of the Treatment Effectiveness review in 1996. The Review clearly stated that 'The Patients' Charter applies to drug misusers as much as to other client groups and...this should not be disregarded simply because they are engaged in an illegal activity.' A new member of staff who arrived in the wake of the organisational changes, commented, 'I was shocked when I got there – the organisation talked users but didn't involve them.... There was a focus on the white male heroin user.'[76] SCODA thereafter employed a drug user and drew up a drug users service charter and in January 1997 it published *Getting Drug Users Involved,* an overview of current arrangements and a guide to good practice in the field.[77] At the same time, the organisation moved to develop a more professional approach for services and service workers, in line with the government's strategy *Tackling Drugs To Build a Better Britain* which had mentioned the need for quality standards and good practice. The organization worked with Alcohol Concern to develop quality standards for the provision of services QuADS (Quality in Alcohol and Drug Services) and a national pilot took place in the late 1990s. Voluntary services were involved in the national pilot because such standards were to become essential in order to obtain statutory funding.[78] This was a very different ethos from the voluntary sector/ residential services focus of the earlier SCODA.

There were further changes in national co-ordination in the early twenty-first century: the merger of SCODA and ISDD to form a new organisation called DrugScope in 2000, and the establishment of the National Treatment Agency (NTA) as a Special Health Authority in 2001. Organizational change was a favoured option for civil servants seeking to make their mark and ISDD had a new chair, Peter Hayes, chair of a hospital trust in the North West, who was also keen for change. In September 1998, a few months after taking on the ISDD job, he advised Council members that Richard Kornicki, a DH civil servant, had

> made it clear that government values ISDD's library and information base BUT he clearly sees SCODA as the senior of the two agencies and said he would like to see the two organisations merge. He said it was also the view of Mike Trace and SCODA's Chief Executive. (It is certainly the view held by Lord Newton too) The reasons given for the merger are that government believes that an organisation of a larger size than either SCODA or ISDD is needed and government feels it is paying too much in funding two organisations whose activities overlap.[79]

The civil servant was also concerned about the future of the recently established Substance Misuse Advisory Service and thought its future could be compromised by the continued existence of competing organisations.

Turbulence in the structure of this 'second tier' organization continued after the merger. The new organisation, named DrugScope, to which Roger Howard was appointed as chief executive, experienced regular financial crises. Putting both organisations together did not ultimately produce a bigger and better organisation. Its media profile was high, but internally the organisation suffered from a lack of clear objectives. In part this was because of wider changes in the drugs field. ISDD was no longer pre-eminent as a provider of information. There were many more organisations producing information and government campaigns, such as FRANK, offered direct advice. Technology and the internet had undermined the role of the ISDD library, once the dominant information resource in the field. SCODA was affected by the merger. Its claim to speak as a representative body and its links with its membership became more attenuated. So 'information' which had been ISDD's rationale and 'the voluntary sector' which had been SCODA's mutated into a blurred area and it became unclear what the new organization was really representing.[80] A review for government in 2007 by the Cordis Bright consultancy recognized the need for a robust second tier Voluntary and Community Sector organization but was unclear what its competence would be.[81]

There was also a new government-based organisation, the National Treatment Agency (NTA). The NTA was one of a number of agencies (NICE-the National Institute for Health and Clinical Excellence was another) set up by government at arms length but with a clear standard setting function. Both NICE and the NTA were characterised as 'intermediary broker organisations' established by government to develop the three way process of broking between research, policy and practice.[82] The National Treatment Agency was announced by Jack Straw as Home Secretary in June 2000 as a joint English initiative between the Home Office, Department of Health and the UK Anti Drugs Coordination Unit. It was established as a Special Health Authority to give it 'operational independence ... in order to become the authoritative national voice on setting standards for drug treatment, commissioning, and provision.' But as Annette Dale Perera, Director of Policy and Practice at DrugScope, commented in an interview in 2001, 'I think we should be clear that the NTA is a political intervention to enhance cross-government departmental coherence around drug treatment and raise standards and consistency across the country ... it will clearly operate in a framework set by ministers.'[83]

As part of this function the NTA incorporated initiatives which had previously characterised SCODA's new role of national co-ordination. This change was symbolised by Dale Perera's own move to the new organisation as Director of Quality in the autumn of 2001. She brought with her the work on quality standards which SCODA/DrugScope had been developing as QuADS and which the NTA published as Models of Care in 2001. Henceforward work on good practice was located in the NTA. One observer saw the 'professionalisation of the field – it's not the same any more ... It's protocol and guidance driven. A lot of people were not very well trained. Now there are national standards – DANOS – NVQs in care and in management....'[84] This was a very different ethos from the anarchic picture of the 1970s.

The new organisation also incorporated users into its work, so the work SCODA had begun also transferred into the Agency. How to develop user representation initially caused problems. Mike Trace and Sally Taylorson were responsible for user involvement at the start of the new agency, but experienced problems in mediating with the 'user field'. What later became seen as a 'strategic misjudgement' was made.[85] They needed a nationally funded group to encourage the development of user groups round the country and decided to work with Matt Southwell of the National Drug User Development Agency (NDUDA), rather than with the Alliance, which represented a more 'medicalised' model. NDUDA was to produce a report: 'There was a first draft and then it went quiet.'[86] The charity Comic Relief was also involved in funding the NDUDA, whose organisational failure led to problems for the team at the NTA.[87] They had to spend time organising events themselves, a process which Taylorson described as involving 'mountains of eggshells'. Some people had their own drug habits and meetings would get forgotten: organising a user friendly meeting was a tense affair.

> It took a long time to get the planning done, but the events were, they said, the best they'd ever come to … 150 hard core users – we had sharps bins in the toilet – in hotels, although not high end, we liaised with local drug agencies and the staff did a sweep of the toilets … with the food, a lot of people were stuffing rolls in bags for later. There were dogs on bits of string and more interesting discussions with security. Lots of users are single parents so we had to make sure child care was in place. Transport – we had to be close to mainline stations and have a shuttle service because of health problems … [88]

One user caused problems by starting to sell *The Big Issue* so Trace hastily defused the situation by buying up the lot. This style of user involvement in one sense epitomised the old style voluntary sector ethos.[89] Trace and Taylorson left the agency in 2002 and the new Chief Executive, Paul Hayes, institutionalised user involvement in a different way, as we discuss in the following chapter. Users featured prominently initially on the Agency's website, but by 2008 their role was more difficult to track down. Peter McDermott was a board member, but by 2008 his status as user representative was not mentioned in his online biography.[90]

The involvement of users in the NTA raised issues of who speaks for the user which will be discussed in the following chapter. The complexities of user involvement were however, also part of wider organisational changes within the first and second tiers of the drug voluntary sector.

This chapter has emphasised the diverse tendencies which characterised the drugs voluntary sector in the early twenty-first century. The rise of the business model took a variety of forms, from the large business style organisations to the social enterprise organisations which still carried forward some of the old voluntary ethos. Alongside these developments had come an erosion of voluntarism in national co-ordination. The NTA had taken over previous roles which the voluntary sector had carried out and typified a process of 'rolling into the state' rather that of the 'rolling back of the state'.

# 7
# Users: Service Users and the Drug User Movement

In April 2006 a group of illegal drug user activists from around the world met at the 17th International Conference on the Reduction of Drug Related Harm in Vancouver, Canada. Together they produced a 'statement about the international network of people who use drugs', a document that is, in effect, an international declaration of drug users' rights. It began:

> We are people from around the world who use drugs. We are people who have been marginalized and discriminated against; we have been killed, harmed unnecessarily, put in jail, depicted as evil, and stereotyped as dangerous and disposable. Now it is time to raise our voices as citizens, establish our rights and reclaim the right to be our own spokespersons striving for self representation and self empowerment.

The statement went on to list a series of key rights which revolved around practical issues such as access to tools which could reduce the harm associated with drug use, such as clean injecting equipment, but also more conceptual goals such as being able to 'survive, thrive and exert our voices as human beings.'[1]

The drug users' declaration, with its employment of terms such as 'voice', 'rights', 'representation' and 'empowerment' was a reflection not simply of specific concerns important to drug users, but of much broader forces at work within contemporary politics and society. The interest of drug users in these issues hinted at the existence of a nascent drug user movement concerned, like other new social movements, with questions of identity and lifestyle. Drug users openly protested about the laws that restrict the use of certain substances and campaigned for the repeal of international regulations that prohibit the taking of drugs as they sought to reclaim drug use as both an identity and a legitimate practice. Indeed, some drug user activists saw their work as part of an international movement, and compared their struggle to the earlier women's movement and the gay rights movement. For one drug user activist there had been a growth of 'activism in all fields, animal rights activists, disabled rights activists, pensioners, old people, ageism activists, feminist activists, gay activists and I suppose the last one to come along, the last one to emerge, is drug user activists.'[2] Activist drug user groups increasingly saw themselves as part of a drug user movement, reminiscent of the new social movements of the 1960s and 1970s and the 'even newer' social movements of the late twentieth and early twenty-first centuries.

However, this was not the only way in which drug users engaged with politics and civil society. In addition to the vocal, campaign-orientated groups there were other types of organisations run by and for illegal drug users. Chief among these were service user groups; groups of drug users who advised service providers at both local and national levels on the design and delivery of drug treatment and other services. Since the 1990s, users of public services were increasingly called upon to take part in discussions about services, their planning and delivery. Growing health consumerism, as well as notions of improving public services through enhanced public engagement, contributed towards a number of measures designed to involve users in public services. In 2001 the Health and Social Care Act was introduced, making it the duty of every organisation providing health and social services to involve individuals using these.[3] As a result, service user groups were established in many areas of public provision, including drug treatment. In the drugs field, service user groups were created locally, regionally and nationally, representing drug users' views to service providers at all levels. These service user groups were 'hybrid' organisations: created by the state but not entirely of it. There were also other groups which represented the interests of 'carers', the friends and families of drug users. These have existed at least since the 1980s and have more in common with 'old' forms of voluntarism, around self-help and mutual aid, but came to assume greater prominence in more recent discussions.

The presence of these related (but distinct) types of 'user group' – the *service user group*, the *activist user group* and the *carer group* – raises important questions about the nature of public services and voluntary activity in the late twentieth and early twenty-first centuries. The state was willing to listen to the 'voice of the user', but there appeared to be some uncertainty as to who was best able to represent that voice. Groups set up by statutory organisations operated in a different sphere when compared to independent groups established by users themselves, but there could be confusion about who these independent user groups actually represented. This chapter will explore the issue of who speaks for the drug user in contemporary Britain. As discussed in previous chapters, a range of voluntary organisations have existed since the 1960s to speak and act on behalf of the drug user. Drug users themselves did play a part in some of these organisations (such as self-help groups) but their role was often 'hidden' from public view. In the 1990s, HIV/AIDS allowed the user to 'come out' and begin to have an influence on drug policy and practice. Since 2000, their influence seems to have increased, as the proliferation of both service user and activist groups demonstrates.

However, there was obviously more than one 'user voice', and there were also a number of constraints to that voice (or even voices) being heard. Drug users have long been a stigmatised group and many had difficultly articulating their views to health professionals and officials in powerful positions. Moreover, like other service user groups, drug users frequently complained that their views were only taken into account in a tokenistic way – little real change occurred in practice. Despite a strong degree of rhetoric around the notion of 'participation' in public services and more widely within civil society, this seemed to remain on terms dictated by the state. Yet, whatever their effect, the existence of such groups is a measure of the changes since the 1960s. Those who draw on public services, whether these are provided by the statutory or the voluntary sector, have increasingly been reconfigured not as passive recipients, but instead as active users.

## 7.1 Who speaks for the user?

### 7.1.1 The growth of user involvement

The emergence of drug service user groups, and to a lesser extent also drug user activist groups, can be related to the gradual growth of health consumerism and the notion that the public, be they citizens, patients or users, should participate in service planning and delivery. As previous chapters have suggested, the idea that patients were also 'consumers' of healthcare contributed towards the growth of voluntary organisations working in the health field from the 1960s onwards, and, since the 1970s, successive governments have drawn on consumerist ideas in the development of aspects of health policy.[4] But, under New Labour, and especially following the publication of the *NHS Plan* in 2000, the individual patient-consumer was placed (at least on the rhetorical level) at the centre of healthcare reform.[5] A series of policy initiatives were introduced with the stated aim of making services more effective and more efficient by improving both patient choice and patient voice.[6] Questions were raised about the implications of such policies, including the extent to which patients actually wanted more choice in healthcare, but the increased emphasis being placed on choice and voice did bring new opportunities for the development of patient involvement.[7] In 2001, the Health and Social Care Act made it a statutory obligation for health and social services to involve service users in the planning and delivery of services. The passing of this act was a reflection of the belief that individuals were demanding a greater say in their own treatment and in the development of services.[8] Improving the capacity for patient voice was also seen as a way of increasing professional accountability and addressing a perceived democratic deficit in health services in favour of the patient.[9] Additionally, some advocates of greater patient involvement maintained that it would encourage people to take a more active role in their own health and healthcare.[10] In his 2002 report to the Treasury on the future funding of healthcare, Derek Wanless argued that for the NHS to remain financially sustainable patients must become 'fully engaged' with their own health and with health services.[11] This meant that individuals were increasingly being seen as 'co-producers' of health, expected to take responsibility for their health by making more healthy lifestyle choices.[12] Involving patients or services users was thus potentially a way of improving individual health and reducing the financial burden on the health service.

Many of these movements seemed to develop organically, away from the state, politicians and policy makers, and their ideas about patient involvement. Indeed, some commentators argue that the notion of increasing patient participation came from patients themselves. Fran Branfield and Peter Beresford, in their study of user involvement produced for the Joseph Rowntree Trust, argued that 'Movements of service users have taken forward both the idea and practice of user involvement.'[13] Yet, the actual extent to which patient movements helped to bring about state interest in patient involvement is difficult to ascertain. As suggested above, politicians and policy makers had a number of reasons for supporting patient involvement quite apart from satisfying the demands of consumers. It seems more likely that patient/ user movement and state interest in patient participation have combined to produce greater levels of patient involvement and the creation of patient or service user groups.

In the drugs field, this could be seen in the appearance of two main types of drug user group: *service user groups* that were often established by statutory authorities or became incorporated within them, although they may well have been run by drug users themselves; and *activist user groups* that usually operated away from statutory bodies and were interested in a wide range of issues around drugs, not just treatment services. Although there was some cross-over between these groups, a clear distinction existed. One drug user activist commented that

> there are a fair few of the UK activists who are involved in and are interested in prohibitionist issues, but not all of them. Some of them are just there to seek better treatment, to improve the treatment system. And don't see much beyond the, don't really look at the social and political context in which they receive treatment. So its two very distinct things and they don't always sit together very comfortably.[14]

Considering the different forms of drug user group separately brings out some of these tensions, as well as areas of agreement and convergence.

### 7.1.2   Service user groups

As discussed in Chapter 6, involving drug users and their carers has been a priority for the National Treatment Agency. Since 2006 the NTA believed that 'service users should be involved in all key aspects of decision-making in relation to their care and that carers should be involved as fully as is agreeable to the user.' Furthermore, taking the Health and Social Care Act into account, the NTA also stated that 'at national, regional, commissioning and service provider level, users and carers should be actively involved in planning, delivery and evaluating service provision.'[15] Nationally, the NTA instigated a number of policies to involve users. A user representative (Peter McDermott) was on the NTA's board; and there was a dedicated users and carers team responsible for national policy and implementation on user involvement. Users were also closely involved in specific programmes, such as the Opening Doors initiative, which aimed to get more drug users into treatment and reduce waiting times. A key part of this scheme was the 'Experts by Experience' programme, established in 2003.[16] Drawing on the wider notion of the 'expert patient', Experts by Experience was created by the NTA (together with the National Institute for Mental Health in England) and run by McDermott with the intention of drawing on the experience of drug service users to 'equip [other] users and carers with practical skills so that they can feel confident and enabled to participate.'[17] This was followed by other practical tools such as the 'Extending Empowerment' online information kit which aimed to 'help providers give service users and their carers a greater say over every area of treatment.'[18] The NTA and other bodies also sought the views of individual users on treatment services, commissioning an annual survey of user satisfaction with services.[19]

This flurry of activity by the NTA around drug user involvement can be attributed to the wider policy context which required service user involvement. For example, many key NTA documents mentioned the Health and Social Care Act, 2001.[20] But, some individuals within the NTA and the broader policy community around drugs also had a particular desire to involve drug service users more fully. Mike Trace's involvement has already been discussed in the last chapter. Although the National Drug User Development Agency (NDUDA) was initially the vehicle of user involvement rather than the Alliance, Trace reportedly told Nelles that 'I'm not going to help you to make revolution on the government's drug policy and prohibition, but what I will help you do is to be more inside the process and to be listened to.'[21] For the NTA, involving users would also help strengthen service accountability, make services more effective and improve the user experience. Users should, it stated in 2006, 'perform a central role as partners within the treatment system' working with treatment providers to improve information about services, feedback to services and develop user influence on services.[22]

From this national framework, user involvement was intended to filter down to the regional, local and individual levels, especially through the Drug Action Teams (DATs), which were required to involve users in service delivery and planning, demonstrating in annual treatment plans submitted to the NTA how and where they were doing this.[23] Many DATs implemented this by employing a 'service user involvement worker.' This worker was sometimes an ex-drug user, but was more often a 'professional' drugs worker.[24] Most DATs also set up a service user group or supported an existing one.[25] 'Notable' examples of user groups were held up by the NTA in 2004 as models for other groups to work towards.[26] One group seen as a particularly successful was the Oxfordshire User Team, or OUT, which was funded by the local DAT and also by the NTA to support the development of other user groups in the region.[27]

These local service user groups performed a variety of roles. Many worked as advocates for service users on a collective basis, informing the local DAT of users' views on issues such as waiting times, prescription policies, care plans and the like. Other groups ran workshops for users on topics such as preventing and dealing with overdoses and coping with Hepatitis C.[28] Some service user groups were also involved in needle exchange. More established groups, such as OUT, offered training to bodies like the police, conducted research and provided advice to other DATs and service user groups.[29] Some DATs also tried to communicate more widely with service users not necessarily involved in the service user group by holding 'stakeholder days' where users offered their input into treatment services alongside service managers and workers.[30]

According to Glenda Daniels, service co-ordinator of OUT, these various programmes of local user involvement were intended to help improve treatment services, give users quicker access to treatment, guarantee that their human rights were respected and ensure that each individual was able to participate in the direction of their treatment.[31] Indeed, individual user participation was also seen to be vital by the NTA. All patients, according to the NTA, had to have influence over decisions about their treatment. Each service user was supposed to have an individual care plan, which represented an agreement between the user and the service provider over the treatment to be given. User involvement was thus intended to be built into every level of service provision from the national level to the individual.

### 7.1.3   Activist user groups

Running parallel to this framework of state-directed user involvement were a number of self-organised drug user activist groups. These were distinct from service user groups, although there were links between the different types of organisation. Activist groups, for example, were also often interested in issues around service provision, but usually as part of a broader social and political agenda that included questions of identity, rights, empowerment and drug policy. Such groups were usually independent from central government and from local agencies such as the DATs, although some received a limited amount of statutory funding.

A number of attempts were made in the early 2000s to co-ordinate these various activist user groups and build a drug user 'movement'. As we discussed in Chapter 6, the National Drug Users Development Agency (NDUDA) was formed at the beginning of the decade. The NDUDA brought together 45 different activist user groups from across the country to support the development of drug user self-organisation. The development agency wanted to be an 'autonomous voice for drug users and their organisations' and 'champion a broad based social policy agenda', which included questioning the prohibition of drugs.[32] Funded by a £30,000 grant from Comic Relief, the NDUDA also initially had the support of Mike Trace, the Deputy Drugs Tsar.[33] However, by 2003–04 the agency had folded, for reasons that will be discussed in greater detail below. The collapse of the NDUDA was followed by other attempts to form a national network of drug user activists, such as the National Users Network (NUN) the National Users Advisory Group (NUAG) and the UK Harm Reduction Alliance (UKHRA).[34]

As we discussed in Chapter 5, a key area of drug user activist work in this period was advocacy within treatment through organisations such as the Alliance. The Alliance claimed to be as 'user-led' as possible, employing users and ex-users as advocates and volunteers, and was governed by a board made up of drug users and sympathetic professionals, including doctors. But drug user activism also took other, non-group based forms. Drug users, for example, produced their own publications, such as *Black Poppy*, a self-styled 'drug users health and lifestyle magazine' written 'exclusively by drug users for drug users.'[35] *Black Poppy* presented a mixture of harm reduction messages, information on drug users' rights, articles on the science and history of drugs, drug users' stories and experiences, poetry, jokes and cartoons. It first appeared in the late 1990s, and was initially funded by Westminster DAT, but funding was withdrawn after the second issue, and *Black Poppy* was subsequently supported by a grant from Comic Relief.[36] They later relied on other charitable donations, (including one from the Getty Foundation) subscriptions and revenue from advertisements.[37] Other user-produced publications, such as the *Users Voice*, were also in circulation. The *Users Voice* aimed to 'include the voices of our peers into the drug policy debate, and to encourage everyone to advocate for themselves.'[38] The *Users Voice* was more international in outlook than *Black Poppy*, reporting on drug user activism from around the world. It was initially funded by SCODA, the Riverside Mental Health Trust, and the Getty Foundation, and received a small grant from the pharmaceutical company Glaxo Wellcome for their office computer.[39]

Other, non-group based work carried out by drug users included involvement in producing research for and about drug users. Drug users, for example, advised on the trial of prescribing injectable heroin at the Institute of Psychiatry, London.[40] They also participated in conferences within the drugs field, something which offered the 'user perspective' to drugs workers and other interested bodies.[41] Furthermore, some drug users helped to train drugs workers and health professionals, including GPs, raising awareness of issues that were important to drug users.[42]

The motives of these drug user activists seemed to be strongly connected to the notion of empowerment. For one activist drug user activism was about 'I think, we feel, taking your control back.'[43] Asked what the purpose of user activism was another activist replied 'I think its trying to empower people and defeat that negative stigma that comes with being an addict, of destroying people's self-esteem.'[44] Two key themes appear to be particularly relevant here: reclaiming the 'drug user' as a legitimate identity; and asserting their 'rights'. Identity has been a particularly problematic issue for drug users, as this was frequently controlled by outsiders. Drug user activists started from a position where their behaviour – drug taking – had already been labelled and ascribed meaning. Labels such as 'addict' 'junkie' 'problem drug user' 'drug misuser' and 'drug abuser' were put upon drug users by doctors, law enforcement agencies and wider society. Some user activists expressed a desire to reclaim their identity as drug users. One activist argued that by 'saying you've got a disease and illness, it keeps [drug use] medicalised and it keeps it your problem.' This, she asserted, 'makes you very powerless.'[45] Indeed, some user activists appeared to find it difficult to escape from their partially medicalised and stigmatised identity as 'addicts'. Another drug user activist commented that 'I think it's always so hard for people like myself who've been through so many different bloody rehabs or treatments and whatever to be very empowered user activists to be honest. Because we've been pathologised...So we're people who're sick.'[46]

Paradoxically, the stigmatisation that users encountered may actually have brought users together and helped to create a collective identity. Erin O'Mara, editor of *Black Poppy*, wrote that her motivation for starting the magazine came from the 'gradual understanding that I – the Editor – and so many of my mates around me, were being treated punitively and neglectfully at many "treatment" and prescribing clinics.' This was, she stated, 'an "awakening" '.[47] User produced publications like *Black Poppy* and the *Users' Voice* used the term 'we', as did the international statement cited at the beginning of this chapter, which suggests a sense of shared identity. Moreover, users were beginning to see themselves as part of a collective movement. One user activist and writer, argued that to achieve change 'users have to become a movement, a non-fragmented organisation.'[48] For other user activists, this user movement already existed. In interviews with user activists, respondents consistently referred to the 'user movement', and have also often compared their own work to that of other social movements. Another user activist saw 'a lot of parallels between the gay movement, even the women's movement, and the drug users' movement.'[49]

Closely connected to these questions of individual and collective identity were also issues of individual and collective rights. Drug user activists were concerned with exercising their existing rights and in asserting claims to new ones. Much attention focussed on the rights individuals had when receiving treatment for drug addiction. The Alliance produced a 'Service Users Charter' outlining what they saw as nine key rights, beginning with the notion that, 'Treatment is a partnership between patients and their treatment team. People should be actively involved in planning their own treatment, and decisions about which treatment or treatments best meet their need should be taken jointly.'[50] However, other organizations and individuals recognized that while users had rights, they may have had some difficulty in exercising these. Drug user activist Peter McDermott, in his 2003 *Out Patient Treatment for Heroin Addiction: A Service Users Guide to Rights and Responsibilities* produced by the voluntary organisation Lifeline, hinted at some of the limitations to users' rights. The guide stated that, 'You are entitled to good treatment services, but these cost money and in return you will be expected to keep your end of the bargain by turning up, not using on top [using drugs in addition to what was prescribed] and to make some changes in your lifestyle.'[51]

For some drug user activists, this represented a compromise that drug users should not be required to make. According to one activist, drug user activists 'suddenly started talking about their right to use drugs.'[52] Though the legal rights of drug users had long been a key interest of earlier voluntary organisations, as seen with the work of Release in Chapter 2, in recent years drug users sought to educate themselves about their rights. *Black Poppy*, for example, produced a guide to users' legal rights entitled 'Drugs, you and the blue light' which told users what their rights were when arrested.[53] But at the same time, user activists also attempted to broaden the notion of legal rights around drugs to argue that they had a human right to use drugs, and that as a result laws prohibiting the use of certain substances should be removed. A worker at the drug policy reform organisation, Transform, initially argued that they believed that the first reason why drugs should be legalised was because 'it's your right to put into your body what you want.'[54] In an article for *Black Poppy* drug user activist Gary Sutton commented that 'Many users would like to see HR (Harm Reduction) stand equally for Human Rights which state our abhorrence of any arm of the state interfering with our chemical needs or pharmacological pleasures.'[55] This perceived right to use drugs led many user activists to argue that 'the users' movement has to challenge prohibition. If it doesn't, it's incomplete. It's just a service lobby.'[56]

For many drug user activists, the notion of human rights included the right to use drugs, and also the right to be treated equally, with dignity and without discrimination. In an editorial for *Black Poppy*, O'Mara contended that 'we believe we are not merely a matter for the shrinks or the courts, but the primary issue is really about Human Rights.' The magazine, she wrote 'was born out frustration, anger and a desire for justice and the right to be treated equally. As the feminist movement said; equal in voice, equal in law, equal in respect – that is equality. As citizens we must be permitted dignity.'[57] Portraying drug use as a human rights issue was also a way of potentially gaining wider support for drug users and the issues that they were concerned with. One user activist spoke with enthusiasm about being able to get Amnesty International interested in the human rights of drug users following Thailand's attempts to curb illegal drug use in 2003.[58]

Indeed, conceiving of drug use as a human rights issue clearly expanded the field of action for drug user activists groups. As a result, activist groups seemed to be concerned with a much broader range of issues than those encountered by service user groups. Where there was some convergence between the two types of groups, however, was over the expertise that drug users had, and the potential value of this experience for services and for drug policy. One drug user activist and writer for *Black Poppy* noted that 'when you've spent that long taking drugs you tend to not have acquired too many other skills...but my knowledge of drugs is pretty good, even if I do say so myself.'[59] User activists argued that drawing on that knowledge was useful to services and service providers. The activist and Release worker Gary Sutton noted that

> Sure, it would be glib to suggest that a drug user would make a better drug worker than a non-user or ex-user, but I don't see why we should be excluded when we have particular qualities as individuals, which can complement our own experience in the field.[60]

Other links between the two types of groups occurred at an individual level. Many user activists were also service users and vice-versa. Moreover, both types of group were interested in influencing policy and practice, although they may well have differed over what goals to pursue.

### 7.1.4   Carer groups

Standing somewhat apart from both activist user groups and service user groups were the carer organisations. Parents' groups, such as the Association for Parents of Addicts, (APA, established in 1967) and ADFAM (established in 1984), had long been a feature of voluntary action around drugs. As we discussed in Chapter 6, both of these organisations moved away from their self-help orientated origins and towards service provision on a large scale. Other groups, such as Families Anonymous (established in 1980), maintained a self-help approach, encouraging the families of drug users to adopt a 'tough love' approach to the drug use of their loved one.[61] These older groups were joined more recently by carer organisations operating in a similar way to drug service user groups. The Health and Social Care legislation requiring the involvement of service users also indicated that their carers should be afforded a role in planning and developing services. The 2002 updated drug strategy reflected the needs of carers as well as users, stating that, 'Parents, carers and families also need support. They experience the problems that drugs cause at first hand, and have a key role in...helping people with drug problems overcome them.'[62] The NTA defined a carer as 'any person who looks after a person identified as a service user or is directly affected by that person's substance use or service use' and has made specific provision for their involvement.[63] In the NTA *Guidance for Local Partnerships on User and Carer Involvement* the agency asserted that, 'The NTA believes that service users should be involved in all key aspects of decision-making in relation to their care and that carers should be involved as fully as agreeable to the user.'[64]

The caveat inserted at the end of this sentence was revealing. Drug users and their carers did not necessarily agree on what was best for the user. Tensions existed between carer groups and drug user groups. Initially, users and carers formed one group, the Interim Advisory Group, to advise the NTA. But, 'after a brief while it became apparent that the two groups did not sit very well together' and separate organisations were formed. This separation, one activist argued, was caused by

> conflicting agendas. And they, the carers agreed about this. They too wanted to split. It wasn't just the users that wanted to split; it was a mutually agreed arrangement. The two groups just wanted separate spaces. I can't say anything more about it than that, it just didn't work together.[65]

Similarly, a worker in a carer organisation commented,

> Families and users – do you keep them separate or together...The carer movement have been quite vocal about their needs. How do you claim to be the carer of someone who is a drug user?...The NTA decided to have carer and user regional fora. In some they are together and in some they are separated. They've decided to be separate overall now.[66]

The NTA began to recognise some of these issues. In their guidance on *Supporting and Involving Carers* the NTA noted that 'It does not work to tag carers on to the needs of users...Carers and families have their own needs.' Moreover, 'At times, the interests and needs of the two groups coincide, but they are often in conflict.' This conflict was perhaps rooted in the fact that most carer organisations developed out of self-help groups, designed to support the friends and family of drug users in helping get their loved one off drugs. Carers, the NTA suggested, 'tend to focus their attention on the individual's drug misusing behaviour.'[67] This raised the difficult issue of how to reconcile the views of the drug user with those of the people who surround him or her. Added impetus had been given to the voice of families through the ACMD's 2003 report *Hidden Harm: Responding to the Needs of Children of Problem Drug Users.* The ACMD estimated that there were between 250,000 and 300,000 children of drug users in the UK and that 'Parental problem drug use can and does cause serious harm to children at every age from conception to adulthood.' The Council recommended that reducing this harm should become a major objective of drug policy and practice.[68]

Balancing the rights of the children of drug users with the rights of drug users themselves could be something of a challenge. If activist drug users insisted upon the 'right to use', what about the rights of their children not to be exposed to the harm that drug use can cause? In this way, carer and family groups provided not only an alternative voice to that of the drug user, but also acted as a counter-balance to the human rights of the user. The presence of such large range of (sometimes contradictory) voices claiming to speak for the user had implications for the impact that these various groups could have upon policy and practice.

## 7.2  Impact and limitations

Assessing the influence of user groups is itself often problematic. As Bruce Wood has noted in relation to generic patient associations, it is difficult to find performance indicators against which to measure the influence or otherwise of patient groups.[69] Marian Barnes and her colleagues found that user groups for mental health and physical disability service users 'had achieved for themselves, and sometimes for those they sought to represent, a degree of influence that could not have been achieved by individuals. Yet, the groups' influence remained fragile.'[70] Brian Salter, in an overview of the role played by patient groups in health policy, noted the massive growth of patient activism in recent years but also suggested that 'the existence of large numbers of vocal health consumer groups does not, of itself, mean that the policy community will feel obliged to listen to them. A policy window is required.'[71] Other commentators argue that this policy window may have already appeared. Rob Baggott, Judith Allsopp and Kathryn Jones studied what they called 'health consumer groups' which aimed to represent patients suffering from a range of conditions including arthritis, cancer, heart disease, mental health problems and maternal health. They found that two thirds of the groups surveyed could identify at least one instance when they felt that they had influenced policy, albeit often over matters of detail rather than overall policy direction.[72]

### 7.2.1   Service user groups

This picture of small-scale influence was, to some extent, replicated in the drug field. One way in which the impact of user groups of both kinds, but particularly of service user groups, can be measured is by looking at the growing number of drug users in treatment. As noted in Chapter 6, the number of people in treatment increased substantially in the late 1990s and early 2000s: this might suggest that users were more willing to make use of drug services because their needs were being met and their views were being listened too. However, critics would point out that at the same time it was increasingly common for drug users to be coerced into treatment, often as an alternative to prison.[73] Yet, the notion that drug users were more content with the services they received would also seem to be confirmed by a user satisfaction survey carried out by the NTA in 2005. The survey found that the 'vast majority' of users surveyed were satisfied with the treatment they received, that users felt respected, they believed that treatment had made a positive impact on their lives, and that they were happy with the staffing of services and the delivery of treatment. Furthermore, 81 per cent of service users felt that they had contributed towards their care plan and 84 per cent believed that the service took their views into account, although 15 per cent of users surveyed thought that their service discouraged complaints.[74] However, another survey of drug service user views, conducted by the European Association for the Treatment of Addiction (EATA) on behalf of the Audit Commission in 2004, was much less positive. The EATA study found that

> Service users reported that [the] opportunity to influence services is virtually non-existent. They reported being disempowered in most aspects of service delivery. Where service user groups exist in local structures (DATs) many felt that this was "lip service" and that they had little opportunity to influence significantly.[75]

Additional questions about the ability of users to influence service provision were raised by a Joseph Rowntree Foundation (JRF) funded study of drug user involvement in treatment decisions which found that 'most interviewees indicated only a limited understanding of what user involvement might mean in a drug treatment context.' What is more, 'not all clients appreciated or were even aware that there was an opportunity for them to be involved in decisions about their own treatment.'[76] Nor did staff and users agree about what they thought the purpose of user involvement was. Service users tended to argue that they should be involved in treatment decisions because they knew best about their own condition and motivations and also that it was their 'right to have control – over their body and their treatment.'[77] In contrast, staff often saw user involvement as a way of getting users to engage with treatment services and improving client retention in treatment, a key NTA goal. Indeed, staff 'tended to emphasise retention in treatment and reduction in drug use as the primary objectives.'[78] This suggests that there were constraints to the ability of individuals to influence the shape of their own treatment, a finding confirmed by the fact that while the authors observed little overt conflict between users and staff within services there was 'latent conflict', where 'clients mutely disagreed with the service being offered but accepted it as being non-negotiable.'[79] The influence of individual service users over service provision would thus seem to be relatively weak. As the JRF report on drug user involvement noted, 'drug user representation at the level of national policy making remains poor.'[80] Similarly, Sue Patterson and her colleagues found that there was a common perception amongst users that national policy making on user involvement was dominated by a few individuals, 'a clique that alienated current service-users who felt undeserving of participation.'[81]

At the local level, additional questions can be raised about the ability of service users groups to effect change. A key concern raised about service user involvement in a range of contexts was that this was often regarded as being tokenistic.[82] Drug service user groups and drug user activists frequently made the same point with respect to drug user involvement. One service user commented that much user involvement had been 'very patronising, very tokenistic.'[83] Another asserted that there is a 'tendency of the field to want to tick boxes, you know, rather than to genuinely engage in an open and creative manner. It's also about wanting to, kind of like, manage the process.'[84] Yet it is not just user activists that have questioned the usefulness of some attempts to involve users. A drug addiction psychiatrist stated that that a lot of user involvement

> is not very well handled, and it's rather tokenistic. Which is not to say it's not better than nothing... but its got to be sort of disappointing when all you've achieved is a sort of bums on seats achievement, and you haven't actually achieved proper integration and influence in the actual decision making process.[85]

Indeed, Patterson and her colleagues found that some individuals connected with drug services viewed user involvement as a 'privilege and [sought] user input only in specific limited areas.'[86]

This statement echoes work on user involvement in health services more widely, which has suggested that this operated as a 'technology of legitimation', where managers 'played the user card': using user groups to support decisions when it suited their purposes, and ignoring them when it did not.[87] Similarly, Barnes and her colleagues found that while government injunctions to listen to user views had some effect on services, professionals and managers adopted a 'tactical' approach to this.[88] Branfield and Beresford observed that service users often believed that service providers 'are slow to change and reluctant to shift the power imbalance that exists between service users and professionals. Service users repeatedly talked about their feelings of powerlessness and futility.'[89] Drug service users and user activists frequently talked in similar terms. In an interview a user activist discussed some of the problems she encountered in her early attempts to set up a service user group at her drug clinic. During a meeting with the consultant in charge where the activist put some of the users' concerns to him, she said that the doctor 'grabbed the paper off [out] of my hand and said "This is not a democracy. If you don't like it, you can leave."' She felt that 'doctors are very heavy handed about what's theirs and getting it [power] out from under that is gonna be really difficult' but at the same time 'we're going to need their help as well.'[90]

This illustrates the fact that drug service users, like other health service users, were dependent on not only the service provided, but also on the goodwill of those individuals providing it. This could act as a further limitation to user influence. What is more, the condition that users were seeking treatment for also seemed to be an important factor in shaping their ability to influence policy and practice. Baggott, Allsop and Jones found that user groups representing patients with conditions that attracted public sympathy, such as cancer and heart disease, were more likely to have an impact than those working for patients with conditions that met with less public understanding, such as mental health problems.[91] This finding is confirmed by much of the literature on mental health user groups, which has pointed to the stigmatisation and oppression of the mentally ill as a barrier to their ability to effect change.[92] This was also a particular problem for drug users; indeed the stigmatisation that drug users faced may even be more severe than that encountered by mental health service users as drug users were frequently associated with illegal activity.[93] Sociologist and activist Sam Friedman contended that, 'Drug users are the target of widespread stigmatisation, derision, and repression. Drug users' organisations are formed, in part, to resist these attacks and to defend the dignity and rights of drug users.'[94] Yet the stigma attached to drug use could make it difficult for drug users to engage with policy makers and professionals. One user activist said that, 'The discrimination against drug users, you know, is so pervasive, it's everywhere you go.'[95]

Drug user groups did seem to encounter specific difficulties which further limited their ability to influence policy and practice. Many of these problems stemmed from the fact that the possession and distribution of what have been classified as 'dangerous drugs' (such as heroin and cocaine) was illegal. Although laws to control dangerous drugs have been in place since the 1920s, some commentators argue that since the late 1990s British drug policy has become increasingly 'criminalised'.[96] Criminal justice measures such as the introduction of Drug Treatment and Testing Orders (DTTOs) which compelled users to attend treatment as an alternative to other types of sentence, including prison, added a stronger coercive element to British drug policy.[97] Indeed, measures such as DTTOs might actually have worked against the facilitation of greater user involvement, as the user was given less say in his or her treatment. Moreover, the legal controls on drug use may have made it difficult for users to come forward for treatment in the first place, leaving a 'hard to reach' population of drug users not in treatment or known to the criminal justice system who were largely unrepresented.

### 7.2.2  Activist users

The laws that surrounded drugs also created a particular problem for drug user activist groups. For the drug user activist groups that sought to have drugs decriminalised or even legalised the drugs laws were clearly a source of their activism. Activists like Michelle Cave argued that, 'Treatment, while important is not enough': users 'need to address the most important question of all – prohibition.'[98] Yet many groups found campaigning against prohibition difficult. A key issue was funding. As a user activist remarked 'it's like how do you raise money for an organisation that demands the right to take drugs in the privacy of your own home? You know? Somebody would just go "Fine. Go do it. But I'm not giving you money to do it." '[99] User organisations that tried to take on prohibition encountered problems. One of the reasons cited for the collapse of the NDUDA was its attempt to co-ordinate efforts to campaign for the legalisation of drugs: this made it difficult to find financial support for the NDUDA, and government funding was not forthcoming after the initial grant from Comic Relief ran out.[100] It was also hard to reconcile the NDUDA's political aims with charity law, which restricted charities' ability to campaign against existing laws. Furthermore, the NDUDA suffered from internal problems. According to Mat Southwell and Tam Miller, (the Chair and Vice-Chair respectively of the NDUDA) the organisation lacked depth as there was only a limited pool of drug users with 'experience of organisational development or management' that could be called upon to work with the agency. This placed too much pressure on the NDUDA's few staff, a situation worsened by the chronic health problems faced by many drug users.[101]

Other user activists suggested that the NDUDA collapsed, in part, because of financial irregularities.[102] Irrespective of the veracity or otherwise of this, the NDUDA was not universally supported by user activists. One activist asserted that 'whilst he had every respect for people who want to kind of get involved in the legalisation, decriminalisation agenda' he was 'much more interested in doing stuff that improves people's lives in the short term.' Moreover, he felt that 'when you kind of confuse and confound the two issues [improving treatment and repealing prohibition] people get like really pissed off [because] ... they feel as though they're spending public money to fund some kind of political campaign. And that pisses people off. And rightly so, if you want to do that stuff, I don't think you can take public money for it.'[103]

According to Harry Shapiro, editor of the journal *Druglink*, disagreement between user drug user activists resulted in 'a self-destructive fracturing into different alliances and ideologies.'[104]

Infighting and tension within such a movement was far from unique. Indeed, as we discussed in Chapter 6, the problems encountered by the NDUDA strongly paralleled the difficulties encountered by the gay AIDS organisation, the Terence Higgins Trust, and attempts to form a national coordinating body for AIDS organisations almost 20 years previously.[105] Competition for limited amounts of financial support and the pressure placed on a small group of committed user activists were also common problems for user organisations of all kinds.[106] Yet, what the tension between user activists who supported an anti-prohibitionist agenda, and those who advocated a more treatment-orientated programme, also illustrated was that drug users did not all want the same thing. The differences between drug users led some individuals, such as former SCODA worker, to wonder 'how representative of users the users groups are. It's an interesting question.'[107] When asked about the meaning of user representation a drug addiction psychiatrist responded 'do you mean a voice from someone in a methadone programme, a sort of Bill Nelles type voice, the rights of a patient population to have their medication treatment, or do you mean a therapeutic community type user perspective'?[108]

Clearly, there was not one 'user voice' but many 'users' voices'. Drug user activists countered the accusation that their groups were not representative by questioning 'how representative is any group?'[109] One activist argued that, 'Striving for absolute representative democracy in a group like this is just impossible. You're simply not going to get it...what's more important [is] that you have something that's functional and efficient.'[110] Similarly, another activist suggested that 'I think as long as you've got people who have a knowledge of drug issues, if it's a user group, that have experienced drug issues...they don't have to know everything, just as long as you've got a...reasonable spreading of skills.'[111] Addressing such issues was important, as they were vital to the wider significance of not just drug user involvement, but the concept of patient involvement as a whole.

Despite tensions within drug activist groups in the UK, their appearance can be seen as evidence for the development of a wider, global drug user movement. By the beginning of the twenty-first century drug user groups, both service orientated and activist orientated, could be found across Europe, North America and Australia, and also in parts of South America and Asia.[112] The international manifestation of user activism led some commentators to posit the existence of a 'social movement in formation'. A group of Nordic researchers suggested that while drug user groups in a national context often appear 'weak, fragmented and marginalised...the picture is very different if we look at them not as separate and isolated national phenomena, but rather as a part of a broader transnational current.' They contended that, 'The idea of a movement becomes more relevant when the minor associations are considered as part of a more widespread trend that seeks to address, question and even challenge the conditions and policies that define and structure drug users' lives.'[113]

By the mid-2000s users were becoming much more visible on the international stage, particularly at drug policy conferences, such as those organised by the International Harm Reduction Association (IHRA), which had long tried to facilitate drug user involvement in its annual conference through measures such as providing scholarships to enable users to attend, giving users a slot to speak in and arranging for methadone clinics or access to substitute prescriptions to be made available to users going to the conference.[114] IHRA also recently supported the establishment of the International Network of People Who Use Drugs (INPUD), a group which aims to co-ordinate user activism around the world.[115] At the IHRA conference in Vancouver in 2006, user activists were allocated two major sessions to run, and given the space and facilities to organise an international drug user congress. This meeting involved around a 100 people from around the world, and according to an IHRA official 'was slickly organised, they had an agenda, people speaking in turn, which is an achievement. People not speaking just personally in an aggrieved way, but you know, conceptually, strategically.' This was, he stated, a 'Huge achievement'.[116]

User activists from around the world were not only willing to work with each other; they were also increasingly collaborating with other voluntary organisations and NGOs in the drugs field. Various international coalitions, such as IHRA, already existed around harm reduction but new groups and networks were being formed. The European Network for Just and Effective Drug Policies (ENCOD) was a network of NGOs that was part of a global group of organisations that supported the Manifesto for Just and Effective Drug Policies. ENCOD regularly lobbied the United Nations (UN) Commission on Narcotic Drugs (CND), they petitioned the European Union Drugs Summit in 2004 and spoke at the IHRA conference in 2007.[117] Other organisations were also trying to lobby the UN and the CND to introduce more harm reduction measures, and as part of this, greater user involvement. A group of activists from IHRA, Human Rights Watch, Canadian HIV AIDS Legal Network, the Beckley Foundation and the Open Society Institute tried to gain access to CND meetings, which could be difficult.[118] To be represented at the UN, NGOs must have ECO-SOC consultative status; that is they were recognised by the UN Economic and Social Council (ECO-SOC) and allowed to participate in meetings. Obtaining this accreditation was a lengthy process, and even when organisations have ECO-SOC status a member country may object to the presence of NGO representatives at a particular meeting.[119] These were just some of the barriers that all NGOs operating at an international level faced, but drug users' groups encountered particular problems. As Anker and his colleagues noted, drug users' organisations 'are often much more introvert, defensive and vulnerable than the powerful collective actors that are traditionally described as social movements.'[120] Clearly, the development of an international drug users' movement should not be overstated, as there were many limitations to the power and influence of drug user groups at all levels.

## 7.3  Conclusion

Indeed, the experience of drug user groups of both the activist and service orientated type raises a series of bigger questions about the concept of user/patient involvement and also the shape of voluntary organisations in the twenty-first century. Drug service user and activist user groups appear to have had a fairly limited impact on service provision and on drug policy. Yet, the emergence of such groups should not be dismissed as irrelevant, chiefly because their presence illustrated important changes in the relationship between the state, voluntary organisations and the individual. The appearance of service user groups was indicative of the growing interest of the state in the views of the user – in order to make services more efficient and more effective – and also to involve the user in the production of their own health. This suggests a reconfiguration of citizenship, where the individual was expected to participate not only in their own treatment but also in the overall direction of services and service provision. However, there appeared to be clear limits to how far state was prepared to listen to views of user. This occurred at the level of services, but also at the broader policy level.

In the drugs field, this could be seen around supervised consumption of substitute drugs – as discussed in Chapter 5. The Alliance was able to influence the implementation of a policy but not its overall shape. As Salter has argued 'the translation of consumer pressure into significant power shifts in the health policy community is a complex process. Visibility in the political arena is no guarantee of influence over the policies produced or the resources allocated.'[121]

The interest of the state in user involvement was also paralleled by an apparent growth in user activism from 'below'. Users came together to form their own groups with aims and objectives that did not necessarily match those of government. This suggests that the notion of user involvement was not just a concept developed by the state, but something that also came from users themselves. This type of user activism would seem to have more in common with traditional models of voluntary organisations, where groups of individuals came together to attempt to solve a common problem. In contrast, service user groups were hybrid organisations: set up either by the state or in accordance with its wishes, but drawing on voluntary effort. There are parallels here with what happened to voluntary organisations working around AIDS in the 1980s and 1990s. Yet, there is also an important difference between the earlier gay AIDS organisations and drug user groups: rather than seeing a transition from 'pure' voluntary groups towards more state funded voluntarism, this latter style of organisation was already present. The significant role played by the state in creating service user groups points to important shifts in the nature of voluntarism and also the relationship between the voluntary sector and the state in the twenty-first century; issues that will be explored in the Conclusion to this book.

# Conclusion

In his Foreword to the 2007 report *The Future Role of the Third Sector in Social and Economic Regeneration* the Prime Minister, Gordon Brown, set out his vision 'of how the state and the third sector working together at all levels and as equal partners can bring about real change in our country.' Brown asserted that, 'In every part of our society voluntary organisations, community groups and social enterprises are making people's lives better, are fighting inequality and making a better environment for us to live in.' Furthermore, 'The third sector is also helping to change the way we think about business. Thousands of social enterprises are changing the decisions we make as consumers and delivering social and environmental outcomes using business approaches.' Taken together, the measures contained in the report were intended to set out a 'new framework for a growing partnership between the third sector and Government.'[1]

In many ways, Brown's Foreword neatly encapsulated some of the key changes that have taken place around voluntarism since the 1960s. The Prime Minister's comment pointed to a drawing together of the 'voluntary and community sector' or, as it is increasingly known, the 'third sector', with the state and the market. Although these three entities had long played a role in the provision of health and welfare services in Britain, by the early twenty-first century the degree of overlap between these areas had increased. The emergence of social enterprises, which combine business models with a voluntary sector ethos, but are often recipients of government funding, exemplified such a trend. In the drugs field, organisations like Exchange Supplies did not sit easily in any category, existing somewhere between the state, the voluntary sector and the market. The use of professional workers and government money by voluntary groups, for example, meant that it was necessary to think in terms of 'hybrid' organisations.[2]

Yet at the same time, the Foreword to *The Future Role of the Third Sector* also hinted at some important continuities. The report spoke of a 'partnership' between the third sector and the state, a term that has been used since at least the beginning of the twentieth century to describe the ways in which the state interacted with voluntary organisations.[3] However, the persistence of a rhetoric of partnership does not necessarily mean that the relationship between the voluntary sector and that the state has remained the same. A number of commentators argue that in recent years the state has assumed a much more dominant role, seriously undermining the independence of the voluntary sector. The writer Nick Seddon, in a report for the think-tank Civitas, entitled *Who Cares? How State Funding and Political Activism Change Charity* argued that, 'While the New Labour government, like the Conservative government that preceded it, talks enthusiastically about independence, this is in truth a concept that is assailed even as it is affirmed.'[4] Seddon suggested that organisations which receive a significant proportion of their income from the state quickly lose their independence. He contended that 'the simple fact is that he who pays the piper gets to choose the tune...Charities slowly but surely find that they are following government agendas.' According to Seddon, organisations which are dependent upon statutory funding, such as Barnardos, which receives 78 per cent of its funds from the state, Leonard Cheshire, (88 per cent of funds from the state) and the drug, alcohol and social care agency Turning Point, (96 per cent of funds from the state) are really 'parastatal' organisations, or *'de facto* state agencies'.[5] Similarly, Frank Prochaska has also suggested that 'Those charities that work closely with local or central government will shape their priorities to suit available grants and often forfeit their role as critics of government policy.' Prochaska famously contended in his evidence to the Deakin commission in 1996 that charities were 'swimming into the mouth of the Leviathan.'[6]

However, other analysts of the recent changes in voluntary-statutory relations have been less convinced of a wholesale takeover of the voluntary sector by the state. In response to Prochaska, Deakin himself observed that 'Government is no longer – if it ever was – a single marine monster but a shoal of smaller beasts, some of which have no ambition to devour large voluntary bodies.'[7] Deakin contends that most voluntary organisations are capable of understanding that limits must be set in order for them to retain their voluntary sector values and status. This is supported by the findings of the 2007 CENTRIS report on the value and independence of the voluntary sector, written by Barry Knight and Sue Robson. Their survey of 121 different voluntary organisations discovered that, 'Most organisations feel themselves to be independent, and although it is clear that public funding does diminish the sense of an organisation's independence, freedoms are diminished at the margin, rather than in a wholesale way.'[8]

## C.1 Voluntarism and the state

The growth of voluntary action around illegal drugs over the last 40 years provides a useful lens through which wider developments in the relationship between the state and civil society can be examined. When the first drug voluntary organisations began to appear in the 1960s and 1970s, many were offered or sought out statutory support. For some organisations, such as Release, the decision to accept government funding was difficult. They feared a loss of independence, that statutory funding would necessarily compromise their ability to criticise the government's drug policy. Yet the organisation was allowed to continue with their sometimes controversial work, such as the drug helpline, without government interference. Similarly, the information and research organisation ISDD was initially wary of statutory support lest it compromise their reputation for independent advice. But, the ISDD's director, Jasper Woodcock, found that he was able to resist Department of Health attempts to influence the allocation of resources to specific projects. Other groups, such as the Community Drug Project, actually demanded statutory funding as they felt that they were fulfilling an important role which the state was neglecting.

The often reciprocal nature of the relationship between the voluntary sector and the state was further underscored during the 1980s and early 1990s. In this period more money flowed into the drug voluntary sector from government sources in response to the heroin explosion and the appearance of HIV/AIDS. Voluntary organisations played a key role in drawing attention to these crises and some groups were also involved in shaping government policy and practice. The Department of Health made a concerted effort to involve voluntary organisations as they were believed to be more innovative and effective than statutory organisations. Moreover, the department was also concerned to break the 'stranglehold' of London-based drug psychiatrists on drug treatment by encouraging the voluntary sector, thus locating the response to drug use away from purely medical specialists. This significant level of support for voluntary organisations could be seen as a reflection of wider moves to 'roll back the state' in this period, but such an analysis ignores the strong degree of central control retained by the Conservative government.

A similar pattern of managed independence was seen around the response to HIV/AIDS and drug use when self-help groups filled a vital gap in statutory services which were often slow to meet the challenges posed by HIV and drug use. But, as state provided services expanded, volunteers became professionals. Voluntary organisations working around drugs and AIDS increasingly relied on statutory funds, leading some commentators to argue that such groups had been absorbed by the state.

Fears of incorporation were further heightened by changes in the way money was actually passed from the state to voluntary organisations during the 1990s. Unconditional grants were gradually phased out and replaced by contracts to provide particular services. Some voluntary organisations feared for their independence as contracts made it difficult to openly criticise government policy. But, this relationship can perhaps be better categorised as one of mutual dependence rather than of reduced independence for the voluntary sector. Voluntary organisations gained funding and support, and the state gained the innovative and effective services it was unable or unwilling to provide. Moreover, this relationship was not unidirectional. Some voluntary organisations were becoming more like the state, but in other areas the state was becoming more like voluntary organisations. For example, the establishment of the National AIDS Trust (NAT) as a voluntary sector organisation and the later creation of the National Treatment Agency (NTA) within the statutory sector, illustrate such a tendency. The NAT in its early days was essentially a state organisation in voluntary sector clothing, while the NTA, as an overt state organisation, occupied the terrain previously considered the province of the voluntary sector and the medical profession. At the same time, other voluntary organisations were moving in a different direction. Some voluntary organisations began to behave more like businesses, as the rise of what were said to be the 'big four' service providers in the drugs field (Turning Point, Cranstoun, Addaction and Phoenix Futures) discussed in Chapter 6 demonstrated. The degree of overlap between the voluntary sector, the state and the private sector increased as all three sectors moved closer together.

Yet, there is much that remains distinctive about each, and particularly about the voluntary sector. Interest in voluntary organisations took on a different dimension in the early years of the twenty-first century as political attention was drawn to their potential role in the revitalisation of civil society and in democratic renewal. Civil society, as Jose Harris points out, is an ancient concept although it largely disappeared from political discourse for much of the twentieth century until the publication of Jurgen Habermas's *The Structural Transformation of the Public Sphere* and Ralf Dahrendorf's *Society and Democracy in Germany* in the 1960s. The translation of these texts into English during the late 1980s coincided with the end of the cold war, and re-invigorating civil society came to be seen as a way to re-build ex-communist and conflict torn societies.[9] More recently, academic and political interest in civil society and in social capital (seen as essential for the maintenance of a strong civil society) has been transferred into Western European and North American contexts. The work of American political scientist Robert Putnam has been particularly influential. Putnam, in an article (and later book) entitled 'Bowling Alone' argued that over the course of the last 30 years Americans had become less engaged in their communities and less connected to each other: that they had become more inclined to bowl alone, rather than together in leagues. This was, he argued, because of a decline in social capital. Social capital can be defined in many different ways, but for Putnam it revolves around the notion that social networks have value, that connections between individuals through norms and networks create reciprocity and trust which is beneficial to all.[10] Although Putnam observed a decline in social capital and, therefore, the demise of American community since the 1960s, he contended that this trend could be reversed if social capital were created or re-created: civil society would be renewed. Putnam recognised that this would be a difficult task, but could be achieved by fostering a greater sense of community engagement and civic participation, and one way of achieving this was through voluntary and community organisations.

Although questions have been raised about the application of Putnam's thesis to Britain, the notion of 'social capital' was enthusiastically taken up by the New Labour government.[11] A review produced by the Office of National Statistics (ONS) in 2001 found that 'social capital' had firmly entered the political lexicon with the ONS, the Home Office, the Department for Education and Skills and the Department of Health all carrying out research to measure and develop different aspects of social capital. Evidence was produced which suggested that high levels of social capital resulted in lower crime rates, better health, greater educational achievement and economic growth.[12] For this reason, what was often termed 'civil' or 'civic renewal' became a key government objective, with policies largely being directed towards the encouragement of volunteering and community self-help as a way of generating social capital.[13] This can be seen, for example, in the 2002 Cabinet Office report *Private Action, Public Benefit*, where the government asserted that a central element of its strategy for the voluntary sector was to help 'charities and other not-for-profit organisations play a bigger role in revitalising communities and empowering citizens.'[14] This notion of empowering citizens was further developed in *The Future Role of the Third Sector in Social and Economic Regeneration*, which devoted an entire chapter to 'Enabling voice and campaigning'. The report asserted that, 'The role of the third sector in providing a means for individuals and communities to make their voices heard and in promoting those voices to campaign for change is critical in supporting civic renewal.'[15]

## C.2   Voluntarism, drugs and the user

Despite this outward government commitment to supporting the campaigning role of voluntary organisations as a way of strengthening civil society, evidence from the drugs field suggests that this policy encounters difficulties on the ground. The appearance of both service user and activist user groups seems to fit quite well with the notion of reviving civil society through voluntary organisations as these groups gave voice to previously disenfranchised individuals often excluded from other means of democratic and political participation. But, as this book has shown, there were a number of limitations placed on the 'voice of the user' undermining their ability to be heard and effect change. Indeed, the entire civil society debate has been oddly absent from the drugs field. Instead, there have been two recent shifts which seem to be working in the opposite direction; adding additional constraints to the 'voice of the user'. These are first, the revival of abstinence within drug treatment policy and practice; and second, an increase in the number of criminal justice measures put in place to deal with drug use. The presence of these constraints raises further questions about the depth of government commitment to the generation of different forms of social capital, and also to polices which purport to engage the recipients of public services in meaningful discussion.

Following the appearance of HIV/AIDS amongst injecting drug users in the late 1980s, a putative consensus formed around harm reduction and methadone maintenance as a way of avoiding the spread of HIV.[16] Oral methadone maintenance became the main method of treatment offered to drug users so that by 2003–4, 63 per cent of patients in structured drug treatment were receiving substitute prescription, and 75 per cent of these were receiving oral methadone on a maintenance basis.[17] Abstinence appeared to drop off the drug policy agenda, and did not feature in either the 1998 or 2002 drug strategies.[18] Although the strength of the supposed consensus around harm reduction and particularly methadone maintenance can be exaggerated, public opposition to this in the late 1990s and early 2000s was rare. Yet, in 2009 as this book was being finalised, there were signs that this situation was changing. A heated debate on the effectiveness or otherwise of methadone maintenance at the National Drug Treatment Conference in Glasgow in 2006 had been followed by an article in *Drink and Drug News* entitled 'Methadone challenged on its home turf: is there a worrying methadone backlash about?'[19] Doctors Tom Carnwarth and Chris Ford detected a possible shift away from methadone maintenance and towards 12-step programmes and drug-free residential rehabilitation. Their article had prompted a stream of letters. Dr David McCartney, an Edinburgh GP with a 'special interest in addictions' asserted the majority of drug-addicted patients actually wanting abstinence rather than harm reduction. 'Are we' he contended 'to drown out their voices with a tidal wave of methadone?'[20]

Support for McCartney's argument came from a number of studies conducted on drug users going into treatment. Professor Neil McKeganey and his colleagues from the Centre for Drug Misuse Research at the University of Glasgow found that 56.6 per cent of drug users entering treatment stated that abstinence was the only change they hoped to achieve by entering treatment.[21] The NTA's first annual user satisfaction survey found that 81.2 per cent of heroin users taking part in the survey stated that they would like to stop using the drug completely.[22] There are, of course, problems with such findings, but these have been used by a group dubbed the 'new abstentionists' to make the case for placing greater emphasis on achieving abstention from drugs within drug treatment.[23] What is more, this debate seemed to be picking up momentum as it spilled over from the drug treatment field and into the public eye. In October 2007 the BBC ran a news story asserting that, 'Treatment services in England have made slow progress in increasing the numbers of people they get off drugs, despite a £130m rise in their budget.' Drawing on the NTA's own figures, the BBC contended that fewer than 3 per cent of individuals left treatment drug-free.[24]

For the 'new abstentionists' this was clear evidence that the government's drug policy was failing. And, drug policy, long an area of consensus between the major political parties appeared to be becoming politicised. Conservative party leader David Cameron had long been a supporter of drug-free residential rehabilitation for the treatment of addiction, and publicly pledged greater financial assistance for residential rehabilitation.[25] Cameron's views were supported by the publication of the policy recommendations to the Conservative party by the Social Justice Policy Group, chaired by former leader Iain Duncan Smith, in July 2007. In *Breakthrough Britain: Ending the Costs of Social Breakdown* the group asserted that, 'Maintenance methadone prescribing which perpetuates addiction and dependency has been promoted under current policy while rehabilitation treatment has been marginalised.' Moreover, there has been a ' "colonisation" of the voluntary sector as the third arm of the state in "harm reduction" drug service provision [which] has stifled innovation and holistic services.' Instead, the group recommended that there should be: 'An expansion of third sector provision of "holistic", value-added, abstinence-based treatment, both day treatment and residential.'[26]

Of course, despite the attention being paid to the 'new abstentionists' there was nothing particularly 'new' about abstention or residential rehabilitation. As we discussed in Chapter 1, residential rehabilitation within therapeutic communities came to Britain in the 1960s and was based around much older notions of self-help and mutual aid. Although residential rehabilitation facilities experienced difficulties in the 1990s (see Chapter 5), they survived this period and began to thrive in a new policy context which appeared to be placing more emphasis on abstention from drugs. What this meant for the user appears to be contradictory. On the one hand, abstinence-based treatment reliant on self-help and mutual aid gave users a place in their own treatment and that of others, as seen from the very earliest days of Phoenix House and programmes like it. Furthermore, as the NTA user satisfaction survey's and McKeganey's research indicates, many users wanted to become abstinent from drugs, or at least they said that they did. The increased focus on abstinence might actually have been giving some users what they wanted. As one long-term worker in the treatment sector put it,

> The not for profit sector and the private sector are the later day revolutionaries – they are offering an alternative to medication...They look and feel like what the voluntary sector looked like twenty five years ago...You can't get a Rizla between the voluntary sector and the NHS – drug treatment means drugs...Some of the prolific offenders are voting with their feet – they are criminal first and foremost – they don't want medication, they want abstinence... [There's] ...a market for total abstinence. 'The first one on our street ever to get drug free'[27]

But on the other hand, holding abstinence up as the most desirable goal within drug treatment undermined the position taken by those activist users who sought to reclaim drug use as a legitimate practice and identity.

This picture was further complicated by the growing interaction between drug treatment and the criminal justice system. Patient choice has been at the centre of New Labour attempts to reform the health service, and yet in the drugs field these choices had to be balanced with the demands of the criminal justice system.[28] A number of new 'interventions' were developed which were designed to get drug users 'out of crime, and into treatment.'[29] Through the Drug Intervention Programme (DIP), introduced in 2003, individuals charged with a so-called trigger offence, such as property crime, robbery, begging and drug offences, could be required to be tested for the presence of an illegal drug.[30] Following the introduction of the Drugs Act in 2005, individuals could also be tested for the presence of illegal drugs on arrest, rather than after being charged.[31] Those who tested positive underwent an assessment by a drugs worker. For individuals convicted of a first offence, this was usually followed by a 'conditional caution', and he or she was referred to treatment without further punishment, although the user could be prosecuted if they failed to attend treatment.[32] Repeat offenders could be issued with Drug Treatment and Testing Orders (DTTOs) which compelled the user to enter treatment as an alternative to other types of sentence, usually prison.[33] The DTTO model was further developed with the introduction Drug Abstinence Orders which gave the courts the power to order an individual to abstain from using heroin, cocaine or any other Class A drug for a period accompanied by regular drugs tests. Failing to obey the conditions of the order could result in further penalties, such as imprisonment.[34] Taken together, measures like the DTTOs and DAOs represented the introduction of a greater degree of coercion within drug treatment, a development that would seem to run counter to moves designed to give patients more choice and more rights in other areas of healthcare.

Indeed, these recent shifts have served only to further complicate the vexed issue of who speaks for the user. As this book has shown, before the 1990s voluntary organisations in the drugs field were largely run *for* drug users, but not *by* drug users themselves. Groups established in the 1960s and 1970s were created in order to fulfil a sense of unmet need that was both practical and political. Drug use was one of a series of new problems that seemed to require new solutions, and voluntary organisations emerged to fill this gap. Drug users were sometimes present in these groups, but 'hidden' from public view. The role performed by drug users in Phoenix House, for example, was an integral part of the organisation and its methods, although the users involved were ex-users rather than current users. Yet despite their 'hidden' status, some ex-drug users may actually have been able to exert a degree of indirect influence over policy. A handful of individual ex-users went on to play an important part in the drug voluntary sector and more widely by sitting on bodies such as the ACMD, although this was more likely to be the result of individual talents than as a concerted effort to speak for users. But, as policy networks began to form around drugs, voluntary groups speaking on the behalf of users were increasingly listened to as the government searched for an effective way to deal with a seemingly ever worsening problem.

In the late 1980s and early 1990s, AIDS helped the user to 'come out'. The public health emergency posed by AIDS in injecting drug users created a place for the user in drug treatment and in drug policy, as the views of users were sought in order to prevent HIV from spreading to the wider population. HIV/AIDS also prompted users to establish their own groups. This sense of crisis brought on by AIDS provided users with the opportunity to influence policy and practice, but as the threat from AIDS appeared to diminish, the degree of urgency decreased. The user voice in turn became divided between users and carers with the latter group more prominent in recent policy documents. At the same time, in the 2000s, interest in the user voice seemed to increase. Service user groups were established in every region of the country, but the ability of these groups to actually influence the shape and direction of services can be seriously questioned. As a recent editorial in the *British Medical Journal* on the involvement of patients in health services commented, 'If we truly wish the public to engage in decisions about health and social care we need to distinguish between initiatives that provide opportunities for meaningful input and action and those that amount to little more than an empty ritual.'[35] Drug users may be 'out' rather than 'hidden' as they were in the past, but are they now ignored?

## C.3   Past patterns, future directions

Throughout the period we have covered in this book, it is notable how developments in the drugs field have often pre-figured later changes in voluntarism, in health and social care, and also in patient involvement. The rise of government funded voluntarism was exemplified in the CFI in the early 1980s. DATs and their bringing together of local 'players' were a portent of later local strategic partnerships which sought to bring together local and health authority players along with the voluntary sector. And the user focussed developments which followed in the wake of HIV/AIDS were in place prior to more general patient consumer changes. The drugs field, often seen as a ghetto, separate from broader developments in health and social policy, in this sense has been a pioneer of later development in those fields.

In turn, this dynamism surrounding voluntary action and illegal drugs was shaped by a number of important recurring themes which existed both inside and outside the drug field. As drug use was an important test area for new policies, models and ideas, there was a role for individual pioneers. People like Griffith Edwards were responsible not only for creating new organisations, but also new ways of working and tackling the problems that drug use created. Yet the involvement of Edwards indicated that professionals in general, and medical professionals in particular, were a part of voluntary groups from the outset. What was also significant about the involvement of people like Edwards was his ability not only to innovate, but also to borrow ideas from other national contexts. Although there were significant differences between British and American drug policy and practice during this period, models and ways of working were copied from the US. Phoenix House was an obvious example, but more recently drug user activist groups have also drawn on models of American activism. European influences and connections were also important, particularly around HIV and AIDS. However, these were not the only models: activism and health was increasingly important in the domestic context as well. This could take many forms, some of which were tied closely to the state, as with ASH (Action on Smoking and Health) whereas others were more separate and even directly opposed to statutory activity.

Clearly, there has not been a neat, linear progression from pure voluntarism in the 1960s to 'para-statal' organisations or state incorporation today. Independent, dynamic voluntary action around illegal drugs can be found throughout this period. Sometimes this operated with statutory support, and at other times without. Some organisations borrowed ideas and structures from the private sector, just as others did not. This complicated, confused and often contradictory picture is, in a sense, a key feature of voluntarism in general. Voluntary organisations showed themselves to be adaptable, responding to changes in drug use and in drug policy, and also to wider trends within health policy, politics and society. It is here perhaps that the one of the key values of voluntary action lies: its adaptability and flexibility. This is something that the state has come to realise. In the *Role of the Third Sector in Social and Economic Regeneration,* Gordon Brown asserted that

> I believe that a successful modern democracy needs at its heart a thriving and diverse third sector. Government cannot and must not stifle or control the thousands of organisations and millions of people that make up this sector. Instead we must create the space and opportunity for it to flourish, we must be good partners when we work together and we must listen and respond.[36]

As the potential for a new or an 'even newer' social movement of activist drug users indicates, voluntary action around illegal drugs continues to develop. The direction of these developments, is however, open to question. There appear to be two opposing trends: one moving towards abstinence and tougher measures on drugs, and the other towards liberalisation and the establishment of drug use as a human 'right'. These trends were illustrated by the appearance, side by side, of two articles in a recent edition of the drug journal *Druglink*. One article envisaged a world where drugs had been legalised by 2022; the other pointed to a 'new Puritanism' which would rapidly make drug taking and binge drinking socially unacceptable.[37] It is impossible to say which direction drug policy will move in, but it is certain that voluntary organisations, however reconstituted and reconfigured, will play a key part in whatever is to come.

# Notes

## Introduction

1. 'Foreword' to G.M. Aves, *The Voluntary Worker in the Social Services: Report of a Committee Jointly set up by the National Council of Social Service and the National Institute for Social Work Training* (London: Bedford Square Press, 1969), p. 11.
2. Home Office, *2003 Home Office Citizenship Survey: People, Families and Communities* (London: Home Office, 2004), p. 188.
3. N. Johnson, 'The changing role of the voluntary sector in Britain from 1945 to the present day' in S. Kunhle and P. Selle (eds), *Government and Voluntary Organisations* (Ashgate: Aldershot, 1992), p. 89; HM Treasury, Cm 7189: *The Future Role of the Third Sector in Social and Economic Regeneration: Final Report* (London: The Stationary Office, 2007), p. 6.
4. National Council for Voluntary Organisations (NCVO), *The Voluntary Sector Almanac 2004* (London: NCVO, 2004).
5. National Council for Voluntary Organisations, *The UK Voluntary Sector Almanac 2006* (London: NCVO, 2006), p. 4.
6. Beveridge quoted in G. Finlayson, *Citizen, State and Social Welfare in Britain 1830–1990* (Oxford: Oxford University Press, 1994), p. 328. For the experiences of 36 voluntary organisations founded in the 1960s, see H. Curtis and M. Sanderson, *The Unsung Sixties: Memoirs of Social Innovation* (London: Whiting & Birch, 2004).
7. M. Gorsky and J. Mohan, *Don't Look Back? Voluntary and Charitable Finance of Hospitals in Britain, Past and Present* (London: Office of Health Economics and Chartered Accountants, 2001).
8. Committee on Voluntary Organisations, *The Future of Voluntary Organisations: Report of the Wolfenden Committee* (London: Joseph Rowntree Memorial Trust, 1978), pp. 22–27.
9. NCVO, *The UK Voluntary Sector Almanac 2006*, p. 8.
10. National Council for Voluntary Organisations, *Voluntary Action: Meeting the Challenges of the 21st Century* (London: NCVO, 2005), p. 8.
11. F. Prochaska, *Christianity and Social Service in Modern Britain: The Disinherited Spirit* (Oxford: Oxford University Press, 2006), pp. 165–66.
12. On the state voluntary relationship in the early twentieth century, see G. Finlayson, 'A moving frontier: voluntarism and the state in British social welfare, 1911–1949', *Twentieth Century British History*, 1, (1990), pp. 183–206.
13. Foreword by Gordon Brown to HM Treasury, *The Future Role of the Third Sector, p. 5.
14. Phrase 'terminological tangle' from Salamon and Anheier (creators of the Johns Hopkins structural-operational definition of voluntary organisations), quoted in N. Deakin, *In Search of Civil Society* (Basingstoke: Palgrave, 2001), p. 9.

15. See http://www.cabinetoffice.gov.uk/third_sector/about/_us, accessed 29 August 2007.
16. Deakin, *In Search of Civil Society*, p. 9.
17. S. Kuhnle and P. Selle, 'Government and voluntary organisations: a relational perspective', pp. 1–33 in S. Kuhnle and P. Selle (eds), *Government and Voluntary Organisations: A Relational Perspective* (Aldershot: Ashgate, 1992), p. 6.
18. See, for example, M. Daunton (ed.), *Charity, Self-Interest and Welfare in the English Past* (London: UCL Press, 1996) and D. Owen, *English Philanthropy 1660–1960* (Cambridge, Mass.: Harvard University Press, 1965). The long history of charity will be discussed in greater detail below.
19. J. Kendall and M. Knapp, *The Voluntary Sector in the UK* (Manchester: Manchester University Press, 1996), p. 10; S. Kuhnle and P. Selle, 'Government and voluntary organisations: a relational perspective', pp. 1–33 in S. Kuhnle and P. Selle (eds), *Government and Voluntary Organisations: A Relational Perspective*, (Aldershot: Ashgate, 1992), p. 6.
20. Deakin, *In Search of Civil Society*, p. 9; N. Crowson, M. Hilton and J. McKay (eds), *NGOs In Contemporary Britain: Non-State Actors in Society and Politics Since 1945* (Basingstoke: Palgrave, 2009).
21. These are discussed in Chapter 2, Chapter 7 and the Conclusion.
22. L. Salamon and H. Anheier, *Defining the Nonprofit Sector: A Cross National Analysis* (Manchester: Manchester University Press, 1997), pp. 33–34.
23. J. Sheard, 'From lady bountiful to active citizen: volunteering and the voluntary sector' in J. Davis Smith, C. Rochester and R. Hedley (eds), *An Introduction to the Voluntary Sector* (London: Routledge, 1995), pp. 115–16.
24. D. Billis, *Organising Public and Voluntary Agencies* (London: Routledge, 1993), pp. 163–56.
25. J. Kendall and M. Knapp, 'A loose and baggy monster: boundaries, definitions and typologies' in Davis Smith, Rochester and Hedley (eds), *An Introduction to the Voluntary Sector*, p. 72; Deakin, *In Search of Civil Society*, p. 9.
26. W. Beveridge, *Voluntary Action: A Report on Methods of Social Advance* (London: George Allen & Unwin, 1948).
27. J. Davis Smith, 'Philanthropy and self-help in Britain 1500–1945' in Davis Smith, Rochester and Hedley (eds), *An Introduction to the Voluntary Sector*, pp. 9–39.
28. *Ibid*, pp. 11–12; M. Taylor and J. Kendall, 'The history of the voluntary sector' in Kendall and Knapp (eds), *The Voluntary Sector in the UK*, pp. 29–33.
29. Smith, 'Philanthropy and self-help', p. 29.
30. *Ibid.*, p. 14; F. Prochaska, *The Voluntary Impulse: Philanthropy in Modern Britain* (London: Faber & Faber, 1988), pp. 21–40.
31. Taylor and Kendall, 'The history of the voluntary sector', p. 40. Prochaska quoted in Davis Smith, 'Philanthropy and self-help', p. 14.
32. M. Daunton, 'Introduction' to M. Daunton (ed.), *Charity, Self-Interest and Welfare in the English Past* (London: UCL Press, 1996), pp. 10–11.
33. Daunton, 'Introduction', p. 12. See also M. Gorsky, 'Friendly society health insurance in nineteenth-century England' in M. Gorsky and S. Sheard (eds), *Financing Medicine: The British Experience Since 1750* (London: Routledge, 2006), pp. 147–64.
34. See M. Gorsky and S. Sheard, 'Introduction' in Gorsky and Sheard (eds), *Financing Medicine: The British Experience Since 1750*, pp. 1–19.

35. M. Gorsky, 'Hospital governance and popular participation in Britain before the National Health Service', presentation to 'NGOs, Voluntarism and Health: Historical and Contemporary Perspectives Workshop', November 2006, London School of Hygiene and Tropical Medicine.
36. Gorsky and Sheard, *Financing British Medicine*; M. Gorsky, J. Mohan and T. Willis, *Mutualism and Health Care: British Hospital Contributory Schemes in the Twentieth Century* (Manchester: Manchester University Press, 2006).
37. Gorsky and Mohan, *Don't Look Back?*
38. V. Berridge, 'Punishment or treatment? Inebriety, drink and drugs, 1860–1914', *Lancet*, 364, (2004), pp. 4–5.
39. H.B. Spear, *Heroin Addiction Care and Control: The British System, 1916–1984* (London: DrugScope, 2002), p. 52.
40. Prochaska, *The Voluntary Impulse*, pp. 70–74.
41. J. Lewis, 'The boundary between voluntary and statutory social service in the late nineteenth and early twentieth centuries', *Historical Journal*, 39, (1996), pp. 155–77.
42. Finlayson, 'A moving frontier', pp. 202–4.
43. See, for example, P. Thane, *Foundations of the Welfare State* (Essex: Longman, 2nd edition, 1996); R. Lowe, *The Welfare State in Britain Since 1945* (Basingstoke: Palgrave, 3rd ed. 2005).
44. Johnson, 'The changing role of the voluntary sector'; Taylor and Kendall, 'The history of the voluntary sector', p. 51, p. 55; M. Brenton, *The Voluntary Sector in British Social Services* (London: Longman, 1985), p. 21.
45. On the voluntary hospitals before the NHS and the wider role of charitable provision in healthcare since then, see Mohan and Gorsky, *Don't Look Back?*
46. Beveridge, *Voluntary Action*.
47. Deakin, 'The perils of partnership', p. 47; Brenton, *The Voluntary Sector in British Social Services*, p. 26.
48. Ministry of Health, *Report of the Working Party on Social Workers in Local Authority Health and Welfare Services* (London: HMSO, 1959).
49. Owen, *English Philanthropy*, p. 527.
50. N. Deakin, 'The perils of partnership: the voluntary sector and the state 1945–1992' in Davis Smith, Rochester and Hedley (eds), *An Introduction to the Voluntary Sector*, pp. 49–50.
51. Finlayson, *Citizen, State and Social Welfare*, p. 329.
52. Advisory Committee on Drug Dependence, (ACDD) *Cannabis* (HMSO: London, 1968), p. 8.
53. Figures from H.B. Spear, 'The growth of heroin addiction in the United Kingdom', *British Journal of Addiction*, 64, (1969), pp. 245–55; p. 247; G. Stimson and E. Oppenheimer, *Heroin Addiction: Treatment and Control in Britain* (London: Tavistock, 1982), pp. 208–9.
54. Ministry of Health, *Drug Addiction: Second Report of the Interdepartmental Committee* (London: HMSO, 1965), p. 5.
55. Dangerous Drugs Act, 1964; Dangerous Drugs (Prevention of Misuse) Act, 1964; Dangerous Drugs Act, 1965; Dangerous Drugs Act, 1967.
56. Ministry of Health, *Drug Addiction: First Report of the Interdepartmental Committee* (London: HMSO, 1961); Ministry of Health, *Drug Addiction: Second Report of the Interdepartmental Committee*; ACDD, *Cannabis*; ACDD, *Search and Arrest* (London: HMSO, 1969).

57. A. Mold, 'The "British system" of heroin addiction treatment and the opening of drug dependence units, 1965–1970', *Social History of Medicine*, 17:3, (2004), pp. 501–17.
58. J. Habermas, 'New social movements', *Telos*, 49, (1981), pp. 33–37; P. Byrne, Social Movements in Britain (London: Routledge, 1997); N. Crossely, Making Sense of Social Movements (Buckingham: Open University Press, 2002).
59. A. Mold, ' "The welfare branch of the alternative society?" The work of drug voluntary organisation Release, 1967–1978', *Twentieth Century British History*, 17:1, (2006), pp. 50–73.
60. C. Coon and R. Harris, *The Release Report On Drug Offenders and the Law* (London: Sphere Books, 1969), pp. 36–37.
61. 1970/71 figure from Brenton, *The Voluntary Sector in British Social Services*, p. 43; 1976/7 figure from Finlayson, *Citizen, State and Social Welfare*, p. 322.
62. Brenton, *The Voluntary Sector in British Social Services*, p. 84.
63. Finlayson, *Citizen, State and Social Welfare*, pp. 357–60.
64. Deakin, 'The perils of partnership', pp. 54–62; Kendall and Knapp, *The Voluntary Sector in the UK*, pp. 201–5; J. Lewis, 'Developing the mixed economy of care: emerging issues for voluntary organisations', *Journal of Social Policy*, 22:2, (1993), pp. 173–92.
65. For an overview of welfare policy under Thatcher, see R. Lowe, *The Welfare State in Britain Since 1945* (3rd edition, Basingstoke: Palgrave, 2005), pp. 317–27; 350–57.
66. Home Office, *Statistics of Drug Addicts Notified to the Home Office, 1988* (London: HMSO, 1989).
67. Advisory Council on the Misuse of Drugs, *Treatment and Rehabilitation* (London: HMSO, 1982), p. 25.
68. Ministry of Health, *Drug Addiction: Report of the Second Interdepartmental Committee*, 8.
69. S. MacGregor et al., *Drug Services in England and the Impact of the Central Funding Initiative* (London: ISDD, 1991), p. l, pp. 71–74.
70. See also A. Mold and V. Berridge, 'Crisis and opportunity in drug policy: changing the direction of British drug services in the 1980s', *Journal of Policy History*, 19:1, (2007), pp. 29–48.
71. Department of Health, *Working for Patients* (London: HMSO, 1989); J. Butler, *Patients, Policies and Politics: Before and After Working for Patients* (Buckingham: Open University Press, 1992).
72. Lewis, 'Developing the mixed economy', pp. 183–91.

# 1   The 'Old': Self-Help, Phoenix House and the Rehabilitation of Drug Users

1. Private Papers of Professor Griffith Edwards (hereafter PP/GE), File 20, 'Stay on Drugs' Phoenix House Information leaflet, n.d. [1970/1].
2. Advisory Committee on Drug Dependence, *The Rehabilitation of Drug Addicts* (London: HMSO, 1968), p. 3.
3. Mold, 'The "British System" of heroin addiction treatment', pp. 501–17.
4. Spear, *Heroin Addiction Care and Control*, p. 52.
5. Finlayson, *Citizen, State and Social Welfare*, pp. 327–28.

6. S. Smiles, *Self-Help: With Illustrations of Character, Conduct and Perseverance* (London: IEA, 1997, first published 1859).
7. Deakin, *In Search of Civil Society*, p. 31.
8. F. Riessman and D. Carroll, *Redefining Self-Help: Policy and Practice* (San Francisco: Jossey-Bass Publishers, 1995), pp. 1–3.
9. See Curtis and Sanderson, *The Unsung Sixties*.
10. D. Robinson and S. Henry, *Self Help and Health: Mutual Aid for Modern Problems* (Martin Robinson: London, 1977), p. 7.
11. M. Hilton, *Consumerism in Twentieth-Century Britain: The Search for a Historical Movement* (Cambridge: Cambridge University Press, 2003), pp. 194–218; pp. 242–67.
12. See, for example, C.J. Ham, 'Power, patients and pluralism', in Barnard, K. & Lee, K. (eds), *Conflicts in the National Health Service* (London: Croom Helm, 1977), 99–110 and M. Stacey, 'The health service consumer: a sociological misconception', *The Sociological Review Monograph: The Sociology of the National Health Service*, 22, (1978), 194–200.
13. B. Wood, *Patient Power? The Politics of Patients' Associations in Britain and America* (Buckingham: Open University Press, 2000), p. 36.
14. G. Smith, 'The rise of the "new consumerism" in health and medicine in Britain, c.1948–1989' in J. Burr and P. Nicholson (eds), *Researching Health Care Consumers: Critical Approaches* (Basingstoke: Palgrave, 2005), pp. 13–38, p. 21.
15. See, for example, the changing outlook and activity of Multiple Sclerosis groups described in M. Nicholson and G.W. Lowis, 'The early history of the Multiple Sclerosis Society of Great Britain and Northern Ireland: a sociohistorical study of lay/practitioner interaction in the context of a medical charity', *Medical History*, 46, (2002), pp. 141–74.
16. H. Perkin, *The Rise of Professional Society: England Since 1880* (London: Routledge, 1989), pp. 472–519.
17. I. Illich, *Limits to Medicine – Medical Nemesis: The Expropriation of Health* (London: Marion Boyars, 1976); M. Foucault, *Madness and Civilisation: A History of Insanity in the Age of Reason* (London: Routledge, 1st English edition., 1967, this edition, 2001); M. Foucault, *The Birth of the Clinic: An Archaeology of Medical Perception* (London: Tavistock Publications, 1973); T. Szasz, *The Manufacture of Madness* (New York: Harper & Row, 1970).
18. Foucault, *Madness and Civilisation*; Szasz, *The Manufacture of Madness*.
19. N. Crossley, 'The field of psychiatric contention in the UK, 1960–2000', *Social Science and Medicine*, 62:3, (2006), pp. 552–63.
20. N. Crossley, 'R.D. Laing and the British anti-psychiatry movement: a sociohistorical analysis', *Social Science and Medicine*, 47, (1998), pp. 877–89; D. Tantam, 'The anti-psychiatry movement' in G.E. Berrios and H. Freeman (eds), *150 Years of British Psychiatry, 1841–1991* (London: Royal College of Psychiatrists, 1991), pp. 333–47.
21. On the influence of anti-psychiatry on the British mental health users' movement see N. Crossley, *Contesting Psychiatry: Social Movements in Mental Health* (London: Routledge, 2005), pp. 127–43. For the impact of anti-psychiatry on the mental health users' movement in the US, see N. Tomes, 'The patient as a policy factor: a historical case study of the consumer/survivor movement in mental health', *Health Affairs*, 25:3, (2006), pp. 720–29.

22. Crossley, *Contesting Psychiatry*, pp. 126–33.
23. *Ibid.*, pp. 144–63.
24. *Ibid.*, pp. 135–39.
25. N. Rose, *Inventing Our Selves: Psychology, Power and Personhood* (Cambridge: Cambridge University Press, 1996), pp. 2–21.
26. M. Thompson, *Psychological Subjects: Identity, Culture and Health in Twentieth-century Britain* (Oxford: Oxford University Press, 2006), pp. 274–75.
27. A. Marwick, *The Sixties: Cultural Revolution in Britain, France, Italy and the United States, c.1968–c.1974* (Oxford: Oxford University Press, 1998), pp. 310–13; J. Green, *All Dressed Up: The Sixties and the Counter-Culture* (London: Pimlico, 1999), pp. 109–12; pp. 191–201.
28. B. Wells, 'Self-help groups' in I. Belle Glass (ed.), *The International Handbook of Addiction Behaviour* (London: Routledge, 1991), pp. 254–57, p. 254; K. Mäkelä, *Alcoholics Anonymous as a Mutual-Help Movement: A Study in Eight Societies* (Madison: University of Wisconsin Press, 1996), pp. 19–24.
29. B. Rawlings and R. Yates, 'Introduction' in B. Rawlings and R. Yates (eds), *Therapeutic Communities for Drug Users* (London: Jessica Kingsley, 2001), pp. 9–25; p. 10.See also, Mäkelä, *Alcoholics Anonymous as a Mutual-Help Movement*, pp. 25–39.
30. B. Wells, 'Narcotics Anonymous (NA) in Britain: the stepping up of the phenomenon' in J. Strang and M. Gossop (eds), *Heroin Addiction and the British System: Vol 2* (London: Routledge, 2005), pp. 167–74, p. 167.
31. D. Kennard, *An Introduction to Therapeutic Communities* (London: Jessica Kingsley, 1998), pp. 21–22; pp. 84–94.
32. R. Janzen, *The Rise and Fall of Synanon: A California Utopia* (Baltimore: Johns Hopkins University Press, 2001).
33. *Ibid.*, p. 83.
34. M. Kooyman, 'The history of therapeutic communities: a view from Europe' in B. Rawlings and R. Yates (eds), *Therapeutic Communities for the Treatment of Drug Users* (London: Jessica Kingsley, 2001), pp. 59–78.
35. M. Rosenthal, 'Therapeutic communities' in I. Belle Glass (ed.), *The International Handbook of Addiction Behaviour* (London: Routledge, 1991), pp. 258–63, p. 259.
36. M. Rosenthal, 'The therapeutic community: exploring the boundaries', *British Journal of Addiction*, 84, (1989), pp. 141–50, p. 145.
37. N. Manning, *The Therapeutic Community Movement: Charisma and Routinization* (London: Routledge, 1989), p. ix.
38. Kennard, *An Introduction to Therapeutic Communities*, pp. 30–42.
39. *Ibid.*, p. 20. For more on the Northfield experiment see T. Harrison, *Bion, Rickman, Foulkes and the Northfield Experiments* (London: Jessica Kingsley, 2000).
40. P. Campling, 'Therapeutic communities', *Advances in Psychiatric Treatment*, 7, (2001), pp. 365–72, p. 366.
41. Kennard, *An Introduction to Therapeutic Communities*, pp. 21–22; pp. 60–63.
42. For a more detailed exploration of these developments see A. Mold, *Heroin: The Treatment of Addiction in Twentieth Century Britain* (De Kalb: Northern Illinois University Press, 2008).
43. Mold, 'The "British System" of heroin addiction treatment'.

44. Advisory Committee on Drug Dependence, *The Rehabilitation of Drug Addicts* (London: HMSO, 1968), p. 3.
45. Spear, *Heroin Addiction Care and Control*, p. 52.
46. *The Coke Hole Trust 3rd Annual Report, 1975–76*.
47. See D. Warren Holland, 'The development of concept houses in Great Britain and Southern Ireland, 1967–76' in D.J. West (ed.), *Problems of Drug Abuse in Britain: Papers Presented to the Cropwood Round-Table Conference* (Cambridge: Institute of Criminology, 1978), pp. 125–35.
48. Interview conducted by Alex Mold with Griffith Edwards, 30 October 2004; T. Cook, 'It was the bad apple: alcohol recovery project' in Curtis and Sanderson (eds), *The Unsung Sixties*, pp. 97–115, pp. 97–99.
49. Interview with Edwards. See also G. Edwards, 'Relevance of American experience of narcotic addiction to the British scene', *British Medical Journal*, 3: 5562, (12 August 1967), pp. 425–29.
50. PP/GE File 5A, Comments on possible adaptation of Phoenix Program to the English scene by Madeline Malherbe, October 1968.
51. PP/GE File 4A, Memorandum from the Community Drug Project to the Dowager Marchioness of Reading, 'The need for American "Addiction Specialists" in England', 30 December 1968.
52. Mold, 'The "British System" of heroin addiction treatment', pp. 501–17.
53. V. Berridge, *Opium and the People: Opiate Use and Drug Control Policy in Nineteenth and Early Twentieth Century England* (Revised edition, London: Free Association Books, 1999), pp. 262–64.
54. A. Trebach, *The Heroin Solution* (New Haven & London: Yale University Press, 1982), p. 183.
55. Warren Holland, 'The development of concept houses', pp. 125–35.
56. Interview with Edwards.
57. PP/GE File 4, Letter from Mr Dixon-Ward, London Boroughs Association to Edwards, 22 December 1969; The National Archives (hereafter TNA) MH 154/367, HM (67)83, The rehabilitation and After-care of heroin addicts, 15 November 1967.
58. TNA HO 383/369, Letter from Home Office official [no name] to Edwards, 13 August 1969.
59. Interview with Edwards.
60. PP/GE File 7A, *Phoenix House: The Featherstone Lodge Project Annual Report, 1970–71*.
61. *Ibid.*
62. PP/GE File 9B, Featherstone Lodge Project, no date [1970?].
63. PP/GE File 7A, *Phoenix House: The Featherstone Lodge Project Annual Report, 1970–71*.
64. *Ibid.*; PP/GE File 9B, *Information for Professionals* no date [1971].
65. PP/GE File 7A, *Phoenix House Annual Report*, 1970–71.
66. PP/GE 7A, Draft article by Phoenix resident Jackie to be sent to *Drugs and Society*, no date [1971/2?].
67. J. Steele, 'I am a junkie – is that what you like?', *The Guardian*, 5 October 1970.
68. PP/GE File 9B, 'Phoenix House', no date [1969/70?].
69. *Ibid.*

70. A. Ogbourne, 'The first 100 residents in a therapeutic community for former addicts', *British Journal of Addiction*, 60 (1975) 65–76, p. 66; PP/GE File 7A, *Phoenix House Annual Report, 1970–71*.

71. PP/GE File 4, Minutes of Planning Group meeting, 4 July 1973.

72. Ogbourne, 'The first 100', p. 65.

73. PP/GE File 7A, *Phoenix House Annual Report, 1970–71*.

74. PP/GE File 11, The need for American addiction specialists in England, no date [1968/9].

75. Interview conducted by Alex Mold with a psychiatrist connected to Phoenix House, 11 May 2006.

76. PP/GE File 9, 'Phoenix House', no date.

77. Definition of experiential knowledge in Baggott, Allsop and Jones, *Speaking for Patients and Carers*, p. x.

78. On tensions between volunteers and professionals, see J.L. Pearce, *Volunteers: The Organisational Behaviour of Unpaid Workers* (London: Routledge, 1993).

79. PP/GE File 3B, Letter from Leida Yuson to Mrs R Clark, 13 September 1970; PP/GE File 3B, Letter from Denny and Leida Yuson to Mitchell Rosenthal and Frank Natale, 19 July 1970.

80. PP/GE File 9B, Letter from Malherbe to Edwards, 18 August 1970.

81. PP/GE File 6A, Letter from Malherbe to Edwards, 5 March 1970.

82. PP/GE File 6A, Letter from Denny Yuson to Colleagues, no date [March 1970].

83. Interview with psychiatrist connected to Phoenix House.

84. PP/GE File 6A, Letter from Malherbe to Edwards, no date [1970].

85. TNA HO 383/396, Letter from Captain John Brown [Chair of the Featherstone Lodge Project Management Committee] to Hague at the Home Office, 10 May 1971.

86. TNA HO 383/396, Memorandum by MH Hogan, 10 June 1971.

87. Interview with psychiatrist connected with Phoenix House.

88. TNA HO383/397, Note for the Drugs Branch re. Mr Graham Harrison's enquiring about Guardian article on Phoenix House by A. Hague, 9 December 1971.

89. PP/GE File 7A, Letter from Captain John Brown to Leida Yuson, 11 August 1971; Letter from Brown to Edwards, 27 September 1971.

90. Interview with Edwards.

91. PP/GE File 7A, Letter from Brown to J. Jordan, 2 November 1971.

92. Interview conducted by Alex Mold with David Tomlinson, 11 July 2006.

93. PP/GE, File 7A, Letter from Brown to Edwards, 16 March 1972.

94. PP/GE File 7A, *Phoenix House: The Featherstone Annual Report, 1971–72*.

95. PP/GE File 7A, Letter from Faith Miles, Phoenix House to Roy Conn, Inner London Probation and Aftercare Service, 21 November 1972.

96. PP/GE File 7A, Letter from Brown to Conn, 22 November 1972.

97. Interview with Tomlinson.

98. TNA HO 383/398, Letter from RIK Blyth, Principal Medical Officer HMP Brixton to David Tomlinson, Director of Phoenix House, 11 October 1977. See also *New Society*, 27 October 1977, pp. 180–81.

99. TNA HO 383/396, Letter from Brown to PG Spurgeon, Principal Officer H2 Division, Home Office, 1 August 1978.

100. TNA HO 383/396.

101. TNA HO 383/396, Note in preparation for meeting, 4 May 1978.
102. TNA HO 383/396 Letter from Spurgeon to Tomlinson, 23 August 1978.
103. A. Ogbourne & C. Melotte, 'An evaluation of a therapeutic community for former drug users', *British Journal of Addiction*, 72, (1977), pp. 75–82, p. 75.
104. PP/GE File 4, *Phoenix House: The Featherstone Lodge Annual Report*, 1972–73.
105. Ogbourne and Melotte, 'An evaluation of a therapeutic community for drug users', p. 76.
106. TNA HO 383/396, Letter from Edwards to members of the management committee presenting Ogbourne's statistical report, 11 March 1974.
107. PP/GE File 4, Phoenix House: First Progress Review, 1973/4.
108. Ogbourne and Melotte, 'An evaluation of a therapeutic community for drug users', p. 80.
109. Interview conducted by Alex Mold with a user activist, 13 December 2005.
110. Interview conducted by Alex Mold with a user activist, 23 August 2005.
111. Ogbourne and Melotte, 'An evaluation of a therapeutic community for drug users', p. 80.
112. Interview conducted by Alex Mold with SCODA worker 25 February 2005.
113. Interview with Tomlinson.
114. *Ibid.*
115. *Ibid.*
116. See http://www.phoenix-futures.org.uk/index.php, accessed 27 September 2007.
117. See DrugScope News, 21 June 2006, http://www.drugscope.org.uk/news_item.asp?a=1&intID=1332, accessed 25 March 2007. For treatment outcome statistics by treatment modality see Department of Health, *Statistics from the National Drug Treatment Monitoring System (NDTMS) 1 April 2003–31 March 2004* (London: Department of Health, 2005), p. 20.The revival of residential rehabilitation is discussed in further detail in the Conclusion.
118. Brian Wells, "Narcotics Anonymous in Britain: the stepping up of a phenomenon," in Strang and Gossop (eds), *Heroin Addiction and the British System, Vol 2*, p. 168.

## 2 The 'New'? New Social Movements and the Work of Release

1. This chapter reproduces sections from A. Mold, ' "The Welfare Branch of the Alternative Society"? The Work of Drug Voluntary Organisation Release, 1967–1978', *Twentieth Century British History*, 17:1, (2006), pp. 50–73.
2. G. Eley, *Forging Democracy: The History of the Left in Europe, 1850–2000* (Oxford University Press: Oxford, 2002), p. 11.
3. A. Touraine, *The Voice and the Eye* (Cambridge and Paris: Cambridge University Press, Editions de la Maison des Sciences de l'Homme, 1981); A. Melucci *Nomads of the Present* (London: Radius, 1988); J. Habermas, 'New social movements', *Telos*, 49, (1981), pp. 33–37.
4. Habermas, 'New social movements', p. 33.
5. Byrne, *Social Movements in Britain*, p. 2.
6. Crossely, *Making Sense of Social Movements*, p. 150.

7. S. Rowbotham, 'Introduction' to Curtis and Sanderson, *The Unsung Sixties*, p. xi.
8. Crossley, *Making Sense of Social Movements*, pp. 150–51.
9. Eley, *Forging Democracy*, p. 10.
10. P. Byrne, *The Campaign For Nuclear Disarmament* (London: Croom Helm, 1988); J. Mattausch, *A Commitment to Campaign: A Sociological Study of CND* (Manchester: Manchester University Press, 1989).
11. Byrne, *Social Movements*, p. 112.
12. Hilton, *Consumerism in Twentieth Century Britain*, p. 198.
13. Crossley, *Making Sense of Social Movements*, p. 165.
14. C. Calhoun, ' "New social movements" of the early nineteenth century', *Social Science History*, 17:3, (1993), pp. 385–427.
15. Marwick, *The Sixties*, pp. 12–13.
16. Green, *All Dressed Up*, p. xi.
17. *Ibid.*, pp. 162–72; Caroline Coon, 'We were the welfare branch of the alternative society', in Curtis and Sanderson, *Unsung Sixties*, pp. 183–197, p. 187.
18. Marwick, *The Sixties*, p. 789; Green, *All Dressed Up*, p. 191.
19. Green, *All Dressed Up*, pp. 160–172, p. 191. Release has experienced recent difficulties, but is still surviving. For information on the current work of Release see http://www.release.org.uk, accessed 14 January 2010.
20. Green, *All Dressed Up*, pp. 173–201.
21. J. Green, *Days in the Life: Voices From the English Underground* (London: Pimlico, 2nd edition, 1988), p. 31, pp. 52–4; figures from ACDD, *Cannabis*, p. 8.
22. *The Times*, 24 July 1967, p. 5.
23. J. Young, *The Drugtakers: The Social Meaning of Drug Use* (London: Paladin, 1972), p. 150, pp. 147–64.
24. *The Times*, 24 July 1967, p. 5.
25. Coon, 'Welfare branch', pp. 183–84; Interview conducted by Alex Mold with Caroline Coon, 5 October 2004.
26. *Ibid.* See also Green, *All Dressed Up*, p. 189; Green, *Days in the Life*, p. 198.
27. Coon, 'Welfare branch', p. 184; Interview between author and Coon.
28. Green, *All Dressed Up*, p. 190.
29. Coon, 'Welfare branch', pp. 184–85; Green, *Days in the Life*, p. 198; Coon and Harris, *The Release Report*, p. 11.
30. Coon, 'Welfare branch', pp. 184–85; for examples of Bust Cards and early statements of Release's purpose see the Release archive, held at the Modern Records Centre, University of Warwick (hereafter MRC) MSS.171/3/12/12, Bust Cards, 1967–1969 and 'The Purpose of Release, 1967'.
31. Coon, 'Welfare branch', p. 185.
32. J. Freeman and V. Johnson (eds), *Waves of Protest: Social Movements Since the Sixties* (Lanham, Maryland: Rowman and Littlefield, 1999), p. x.
33. T. Buchanan, ' "The truth will set you free": the making of Amnesty International', *Journal of Contemporary History*, 37:4, (2002), pp. 575–97.
34. Ham, 'Power, patients and pluralism', pp. 112–14; Hilton, *Consumerism in Twentieth Century Britain*, pp. 167–218; N. Crossley, 'Transforming the mental health field: the early history of the National Association for Mental Health', *Sociology of Health and Illness*, 20:4, (1998), pp. 458–88.
35. Habermas, 'New social movements', p. 33.

36. Coon, 'Welfare branch', p. 185.
37. MRC MSS.171/5/1/2, Jeremy d'Agapeyeff, *Release: A Progress Report* (Norfolk, 1972), Section IV.
38. *Ibid.*, Section XIII; Rufus Harris interviewed in Green, *Days*, p. 198.
39. MRC MSS.171/5/1/2, d'Agapeyeff, *Release*, Section I.
40. *Ibid.*, Section I. See also Harris in Green, *Days*, p. 198.
41. MRC MSS.171/5/1/2, d'Agapeyeff, *Release*, Section I.
42. MRC MSS.171/3/12/11, Typescript of interview between William Wordsworth and Coon, 1970; MRC MSS.171/3/12/17, Press Release about Release and 'Performance', November 1970.
43. MRC MSS.171/5/1/2, d'Agapeyeff, *Release*, Section VI.
44. *Ibid.*, Section XVI.
45. Johnson, 'The changing role of the voluntary sector', p. 89.
46. Finlayson, *Citizen, State and Social Welfare*, pp. 314–16, pp. 327–28. For an account of 36 organisations established in this period including CPAG and Shelter see Curtis and Sanderson, *Unsung Sixties*.
47. Finlayson, *Citizen, State and Social Welfare*, p. 328.
48. *Ibid*, pp. 308–17, Deakin, 'The perils of partnership', p. 50.
49. Harris quoted in Green, *All Dressed Up*, p. 190.
50. MRC MSS.171/5/1/2, Michael Schofield, Foreword to d'Agapeyeff, *Release*.
51. MRC MSS.171/3/12/17, Release cases – preliminary analysis, 1970.
52. ACDD, *Cannabis*, p. 24; Coon and Harris, *Release Report*, p. 109.
53. Afterword to Coon and Harris, *Release Report* by Michael Schofield, p. 120.
54. Coon and Harris, *Release Report*, p. 36, p. 37.
55. MRC MSS.171/3/12/23, Release: A Fifth Anniversary Policy Statement, 1972. Examples of Release's campaigns on these issues can be found in their publications, particularly their newsletters, *Connection* and *News Release*. See MRC MSS.171/4/7-8, *Connection*, 1972 and MRC MSS.171/4/14-23, *News Release*, 1977–1980.
56. MRC MSS.171/4/14, *Release 67–77: Release Tenth Anniversary Publication*, 1977, pp. 13–14; D. Turner, 'The development of the voluntary sector: no further need for pioneers?' in J. Strang and M. Gossop (eds), *Heroin Addiction and Drug Policy: The British System* (Oxford: Oxford University Press, 1996), pp. 222–30, p. 225.
57. Coon and Harris, *Release Report*.
58. MRC MSS.171/3/12/13, Written statement for the ACDD Search and Arrest Committee, 1969; MRC MSS.171/5/1/2, d'Agapeyeff, *Release*, Section XII; Coon, 'Welfare branch', p. 190.
59. MRC MSS.171/3/19/1, Letter from Coon to Mr Bartholomew, 11 November 1970.
60. Spear, *Heroin Addiction Care and Control*, p. 287.
61. Wood, *Patient Power*, p. 187.
62. On the reaction to the Wootton report see Green, *All Dressed Up*, pp. 185–88.
63. V. Berridge, 'AIDS and British drug policy: continuity or change?' in V. Berridge and P. Strong (eds), *AIDS and Contemporary History* (Cambridge: Cambridge University Press, 1993), pp. 135–56, pp. 139–41.

64. Green, *All Dressed Up*, p. 190.
65. Coon, 'Welfare branch', p. 187.
66. *Ibid.*, pp. 186–87.
67. Steve Abrams interviewed in Green, *Days in the Life*, p. 199.
68. Marwick, *The Sixties*, p. 491.
69. Nicola Lane interviewed in Green, *Days in the Life*, p. 401.
70. MRC MSS.171/4/8, *Connection*, 1972, pp. 25–26.
71. MRC MSS.171/4/14, *Release 67–77*, pp. 2–3.
72. F. Prochaska, *Women and Philanthropy in Victorian England* (Oxford: Oxford University Press, 1980); P. Thane, *Foundations of the Welfare State* (Longman: London, 2nd ed., 1996), pp. 18–19.
73. Davis Smith, 'Philanthropy and self-help in Britain', pp. 9–39.
74. For a summary of the recent debates on philanthropy and charity during the nineteenth century, see Daunton, 'Introduction' in Daunton (ed.), *Charity, Self-Interest and Welfare in the English Past*, pp. 10–11.
75. MRC MSS.171/5/1/2, d'Agapeyeff, *Release*, Section XVI.
76. MRC MSS.171/4/14, *Release 67–77: Tenth Anniversary Publication*, 1977, p. 1.
77. MRC MSS.171/3/12/11, Release cases: preliminary analysis, April–June 1970.
78. Figures quoted in Marwick, *The Sixties*, p. 278.
79. MRC MSS.171/5/1/2, d'Agapeyeff, *Release*, Section III.
80. Green, *All Dressed Up*, p. 125.
81. Des Wilson quoted in I. Williams, *The Alms Trade: Charities, Past, Present and Future* (London: Unwin Hyman Limited, 1989), p. 146. See also E. Ware, '£325 to re-house a family' in Curtis and Sanderson (eds), *Unsung Sixties*, pp. 19–28.
82. Finlayson, *Citizen, State and Social Welfare*, p. 332.
83. Crossely, *Making Sense of Social Movements*, p. 164; P. Bagguley, 'Social change, the middle classes and the emergence of "new social movements": a critical analysis', *The Sociological Review*, 40:1, (1992), pp. 26–48.
84. F. Parkin, *Middle Class Radicalism: The Social Basis of CND* (Manchester: Manchester University Press, 1968).
85. Byrne, *Campaign for Nuclear Disarmament*; S. Cotgrove and A. Duff, 'Environmentalism, middle class radicalism and politics', *Sociological Review*, 28:2, (1980), pp. 333–51.
86. Cotgrove and Duff, 'Environmentalism'.
87. Bagguley, 'Social change and the emergence of new social movements', p. 40; C. Offe, 'New social movements: challenging the boundaries of institutional politics', *Social Research*, 52:4, (1985), pp. 817–68.
88. Crossley, *Making Sense of Social Movements*, p. 164.
89. MRC MSS.171/5/1/2, d'Agapeyeff, *Release*, Section V; MRC MSS.171/3/12/11, Release Staff and Release Family, No date [1970]; List of conference participants and qualifications, in D.J. West (ed.), *Problems of Drug Abuse in Britain: Papers Presented to the Cropwood Conference* (Cambridge: Institute of Criminology, 1978).
90. MRC MSS.171/3/12/30, *The Princedale Trust: Bi-Annual Report on the Work of Release*, 1977.
91. Lowe, *The Welfare State in Britain Since 1945*, pp. 277–78.
92. *Ibid.*, p. 279.

93. T. Evans, 'Stopping the poor getting poorer: the establishment and professionalisation of the poverty lobby, 1945–1995', paper submitted for the Database of Non-Governmental Organisations conference, University of Birmingham, July 2007. Paper reproduced as T. Evans, 'Stopping the poor getting poorer: the establishment and professionalisation of the poverty lobby, 1945–1995' in N. Crowson, M. Hilton and J. McKay (eds), *NGOs In Contemporary Britain: Non-State Actors in Society and Politics Since 1945* (Basingstoke: Palgrave Macmillan, 2009), pp. 147–63.

94. Interview between author and Coon.

95. MRC MSS.171/3/12/12, Release annual expenses, 1969–1970.

96. MRC MSS.171/3/12/11, Release Staff and Release Family, No date [1970]; Coon, 'Welfare branch', p. 191.

97. Coon, 'Welfare branch', p. 191; MRC MSS.171/2/1, End of year accounts, 1972–1975; MRC MSS.171/1/2, Business Meeting and Policy Meeting Minutes, 1971–1972, Minutes of the Meeting of the Princedale Trustees, 21 November 1972.

98. Sheard, 'From lady bountiful to active citizen', pp. 115–16.

99. Billis, *Organising Public and Voluntary Agencies*, p. 163.

100. *Ibid.*, p. 161.

101. MRC MSS.171/5/1/2, d'Agapeyeff, *Release: A Progress Report*, Section XVI.

102. *Ibid.*

103. Green, *All Dressed Up*, p. 190; Coon, 'Welfare branch', p. 188.

104. MRC MSS.171/3/12/11, Application for a grant of £15,000 over three years to finance the running costs of Release from the Rowntree Trust, no date, [1970?]; MRC MSS.171/3/19/1, Letter from Rufus Harris to Michael Levin of PLAY 3 July 1970; MRC MSS.171/3/19/1, Letter from Coon to Eric Clapton, 21 April 1970.

105. MRC MSS.171/3/13/25, Estimated donations, April–May 1973; MRC MSS.171/4/23, *News Release*, 7:1, March–May 1980; MRC MSS.171/3/19/1 Donators and fundraising, 1970.

106. MRC MSS.171/3/19/1, Letter from Harris to Levin, 3 July 1970.

107. MRC MSS.171/1/5, Minutes of staff meeting, 18 April, 1974; MRC MSS.171/1/5, Minutes of meeting 23 October, 1973; MRC MSS.171/3/12/11, Application for a grant of £15,000 over three years to finance the running costs of Release from the Rowntree Trust, no date, [1970?]; MRC MSS.171/3/12/12, Press Release: 'Flash! The last of the great rock premieres to benefit Release', 4 January 1971.

108. MRC MSS.171/3/12/28, 'The Cheque Book', no date; MRC MSS.171/1/5, Staff meeting minutes, 29 March 1974; MRC MSS.171/1/2, Princedale Trustees meeting minutes, 13 June 1972.

109. Coon, 'Welfare branch', p. 191.

110. MRC MSS.171/2/1, End of year accounts, 1971–1975.

111. MRC MSS.171/3/19/1, Letter to Harris from John Profumo at Toynbee Hall, 23 December 1970.

112. MRC MSS.171/5/1/2, d'Agapeyeff, *Release*, Section XV; MRC MSS.171/3/19/3, Letter to J.E. Miller from Release, 18 March 1971.

113. MRC MSS.171/3/19/2, Letter to Coon and Harris from Bernard Simons at Clinton, Davis, Simons and Co. [Release's solicitors] 19 March 1971; MRC MSS.171/5/1/2, Schofield foreword to d'Agapeyeff, *Release*.

114. MRC MSS.171/4/8, *Connection*, 1972, 'Image and ideology: the fund-raising dilemma', p. 25.
115. Finlayson, *Citizen, State and Social Welfare*, p. 350.
116. T. Buchanan, 'Amnesty International in crisis: 1966–7', *Twentieth Century British History*, 15:3, (2004), pp. 267–89, p. 269.
117. Finlayson, *Citizen, State and Social Welfare*, p. 350.
118. MRC MSS.171/4/9, *Release Monthly Newsletter*, May 1974.
119. MRC MSS.171/1/5, Minutes of meeting 21 June 1974. On the establishment of the VSU in 1973 see Kendall and Knapp, *The Voluntary Sector in the UK*, p. 8.
120. Interview with Coon.
121. MRC MSS.171/1/5, Minutes of meeting 3 May 1974.
122. MRC MSS.171/1/5, Minutes of meeting 29 November 1974.
123. MRC MSS.171/1/5, Letter from Coon to Release, 29 November 1974.
124. *Ibid.*
125. Lewis, 'The boundary between voluntary and statutory', p. 158.
126. Kendall and Knapp, *The Voluntary Sector in the UK*, pp. 136–37.
127. *Ibid.* p. 139; Deakin, 'The perils of partnership', pp. 52–56.
128. Lewis, 'Developing the mixed economy', p. 184, p. 190.
129. Interview conducted by Virginia Berridge and Alex Mold with Jasper Woodcock, 10 November 2004.
130. MRC MSS.171/4/15-23, *News Release*, 5:1, March–May 1978, pp. 14–16.
131. MRC MSS.171/4/15-23, *News Release*, 5:2, June–August 1978, p. 7.
132. MRC MSS.171/3/44/1, The Release Drug Hotline, draft press release, 1975.
133. *The Guardian*, 28 July 1975.
134. *The Times*, 19 September 1975.
135. MRC MSS.171/3/44/1, 'End of the line for the drugs "hotline" by aid group', *Paddington Mercury*, 15 August 1975.
136. MRC MSS.171/3/44/1, Letter from B. Cross at the Post Office to Release, 24 October 1975.
137. Crossley, 'Transforming the mental health field', pp. 461–63, pp. 474–80.

## 3   Drug Voluntary Organisations and the State in the 1960s and 1970s

1. Finlayson, *Citizen, State and Social Welfare*, p. 305.
2. PP/GE File 3A, Minutes of CDP main committee meeting 2 March 1970.
3. *Ibid.*
4. DL 32718, *SCODA Annual Report, 1978–79*, p. 14.
5. See Mold, 'The "British System" of heroin addiction treatment'.
6. Aves, *The Voluntary Worker in British Social Services*. The idea that voluntary organisations should 'complement, supplement, extend and influence' statutory provision was put forward by the Wolfenden report. See Joseph Rowntree Trust, *The Future of Voluntary Organisations* (London: Croom Helm, 1978), p. 26.
7. DL 14039, *The First SCODA Report, 1974*.
8. Finlayson, *Citizen, State and Social Welfare*, p. 305.
9. Brenton, *The Voluntary Sector in British Social Services*, p. 36.

10. Brenton, *The Voluntary Sector in British Social Services*, pp. 20–22; pp. 132–39.
11. R. Lowe, *The Welfare State in Britain Since 1945* (Basingstoke: Palgrave, 2005), p. 286.
12. Ennals quoted in Brenton, *The Voluntary Sector in British Social Services*, p. 136.
13. Aves, *The Voluntary Worker in the Social Services*.
14. Brenton, *The Voluntary Sector in British Social Services*, p. 45.
15. Joseph Rowntree Trust, *The Future of Voluntary Organisations* (London: Croom Helm, 1978), p. 26.
16. 1970/71 figure from Brenton, *The Voluntary Sector in British Social Services*, p. 43; 1976/7 figure from Finlayson, *Citizen, State and Social Welfare*, p. 322.
17. Brenton, *The Voluntary Sector in British Social Services*, p. 84.
18. Interview conducted by Alex Mold with Rev Kenneth Leech, 9 November 2004.
19. K. Leech, *Care and Conflict: Leaves From A Pastoral Notebook* (London: Darton, Longman & Todd, 1990), pp. 13–14.
20. K. Leech, *Pastoral Care and the Drug Scene* (London: SPCK, 1970), p. 116.
21. Blenheim Project, *People Adrift: The Work of the Blenheim Project With Young Drifters* (London: Blenheim Project, 1973).
22. N. Dorn and N. South, *Helping Drug Users: Social Work, Advice Giving, Referral and Training Services of Three London Street Agencies* (Aldershot: Gower, 1985), p. 23.
23. *Ibid.*, p. 25.
24. Interview conducted by Alex Mold and Virginia Berridge with Rowdy Yates, 10 March 2006.
25. *Ibid.*
26. PP/GE File 4A, Background note on Camberwell Drug Project by Gerry Stimson, 31 May 1968.
27. See B. Thom, *Dealing With Drink: Alcohol and Social Policy – From Treatment to Management* (London: Free Association Books, 1999), pp. 91–94 and T. Cook, 'It was the bad apple: the Alcohol Recovery Project' in Curtis and Sanderson (eds), *The Unsung Sixties*, pp. 97–115.
28. PP/GE File 19, *CDP Annual Report, 1968–69.*
29. *Ibid.*
30. PP/GE File 19, Proposal for a community centre for drug addicts by the Addiction Research Unit, Institute of Psychiatry, no date [1968].
31. V. Berridge, *Health and Society in Britain Since 1939* (Cambridge: Cambridge University Press, 1999), pp. 34–37.
32. C. Webster, *The National Health Service: A Political History* (Oxford: Oxford University Press, 2002), pp. 54–55.
33. Bennett, 'The drive towards community', p. 327.
34. PP/GE File 4A, Letter from Griffith Edwards to John Brown, 18 July 1968.
35. Berridge, *Health and Society in Britain Since 1939*, pp. 44–45. See also M. Jefferys, 'The transition from public health to community medicine', *Society for the Social History of Medicine Bulletin*, 39, (1986), pp. 47–63.
36. PP/GE, File 4A, List of anticipated questions and proposed answers from a press release on the CDP, no date [July 1968].
37. PP/GE, File 19, *CDP Annual Report, 1968–69.*
38. *Ibid.*

39. *Ibid.*
40. PP/GE File 19, Proposal for a community centre for drug addicts by the Addiction Research Unit, Institute of Psychiatry, no date [1968].
41. PP/GE File 3A, Report on the activities of CDP week ending 21 February 1969.
42. PP/GE File 18, *Community Drug Project Fifth Annual Report, 1974–75.*
43. PP/GE File 18, *Annual Report: Community Drug Project, 1977–78.*
44. *Ibid.*; PP/GE File 18 *Annual Report: Community Drug Project, 1974–75.*
45. Interview with Edwards.
46. PP/GE File 18, *Annual Report: Community Drug Project, 1974–75.*
47. PP/GE File 18, Circular regarding fixing room: copies to trustees, management committee and the DDUs, 9 December 1976.
48. Dorn and South, *Helping Drug Users*, p. 13; pp. 23–25.
49. PP/GE File 19, Letter from Edwards to Mr Shifrin, 7 July 1969; PP/GE File 4A, List of anticipated questions and proposed answers from a press release on the CDP, no date [July 1968].
50. PP/GE File 19, Letter from Captain Brown the Mr Shifrin, 4 July 1969.
51. PP/GE File 19, *Community Drug Project Annual Report, 1968–69.*
52. See A. Mold, *Heroin: The Treatment of Addiction in Twentieth-Century Britain* (De Kalb: Northern Illinois University Press, 2008).
53. Coon and Harris, *The Release Report*, pp. 30–31.
54. PP/GE File 18, *Community Drug Project: Fourth Report, 1972–73.*
55. PP/GE File 19, *Community Drug Project: Third Report, 1970–71.*
56. PP/GE File 3A, Minutes of the CDP Main Committee Meeting 2 March 1970.
57. PP/GE File 19, *Community Drug Project: Third Report, 1970–71.*
58. MRC MSS.171/4/18, *News Release* October/December 1977, p. 12.
59. PP/GE File 18, *Community Drug Project: Fourth Report, 1972–73.*
60. PP/GE File 4A, CDP Minutes of Main Committee Meeting, 20 October 1969; PP/GE File 18, *Community Drug Project: Fourth Report, 1972–73.*
61. DL 14039, *The First SCODA Report, 1974.*
62. MRC MSS.171/4/18, *News Release,* October/December 1977, pp. 13–14.
63. DL 26430, Report of the Conference 'Ten Years After', 1 July 1976.
64. PP/GE File 3A, Minutes of CDP main committee meeting, 2 March 1970.
65. PP/GE File 3A, Summary of replies to questionnaires sent to DDUs, 13 May 1970.
66. PP/GE File 3A, Letter from Dr James Willis, Consultant Psychiatrist, Guys Hospital to Edwards, 30 November 1970.
67. PP/GE File 18, Letter from Griffith Edwards to Alan Ogbourne, 9 February 1972.
68. DL 29050, *SCODA Annual Report, 1976–77.*
69. Bewley, "Conversation with Thomas Bewley," 885.
70. See Mold, *Heroin*, Chapter 3.
71. See, for example, Ogbourne, 'The first 100 residents in a therapeutic community for former addicts'; Ogbourne and Melotte, 'An evaluation of a therapeutic community for drug users'.
72. Interview with Edwards.
73. J. Woodcock, 'The growth of information: the development of Britain's nation drug misuse information resource' in J. Strang and M. Gossop (eds), *Heroin Addiction and Drug Policy: The British System* (Oxford: Oxford University Press, 1994), pp. 304–10.

74. TNA FD 23/1949, Proposal for setting up an Institute for the Study of Drug Addiction and related questions, 1967.
75. Interview conducted by Alex Mold and Virginia Berridge with Jasper Woodcock, 10 November 2004.
76. TNA FD 23/1949, Proposal for setting up an Institute for the Study of Drug Addiction and related questions, 1967.
77. TNA FD 23/1949, Letter from Himsworth to Amory, 4 June 1968.
78. Woodcock, 'The growth of information', pp. 305–7.
79. Interview with Woodcock.
80. *Ibid.*
81. F. Logan (ed.), *Cannabis: Options For Control* (London: ISDD, 1979).
82. Interview with Woodcock.
83. Committee on Voluntary Organisations, *The Future of Voluntary Organisations*.
84. TNA MH 154/433, Letter from Mr Eversfield to Mr Platten, Town Clerk, London Borough of Enfield, 20 December 1971.
85. PP/GE, File 4A, List of anticipated questions and proposed answers from a press release on the CDP, no date [July 1968].
86. PP/GE, File 18, Letter from Martin Mitcheson to Captain Brown, 5 May 1971.
87. PP/GE, File 18, *Community Drug Project: Annual Report, 1970–71.*
88. TNA MH 154/433, Letter from Mr Chadwell, Home Office to Mr Ogbourne, CDP, 21 June 1971.
89. TNA MH 154/433, Note of an office meeting between Dr Sippert and Mr Eversfield (DHSS) and Mr Searchfield and Dr Mitcheson (CDP) 23 April 1971.
90. TNA MH 154/433, Letter from Mr Eversfield to Mrs Lee, 26 April 1971.
91. TNA MH 154/430, Points arising from LBA working party on drug addiction, MH Bruce, Social Work Service Officer, June 1972 and TNA MH 154/430, London Boroughs Association, Report of General Purposes Committee, 22 November 1972.
92. TNA MH 154/433, Letter from Geoffrey Finsburg, Joint Parliamentary Under Secretary of State, DHSS to John Fraser MP, regarding the funding of CDP, 22 October 1982.
93. MRC MSS.171/4/14, *Release 67–77: Tenth Anniversary Publication*, pp. 13–14.
94. *Ibid.*; Woodcock, 'The growth of information', p. 304–7.
95. Woodcock, 'The growth of information', p. 307.
96. Interview with Woodcock.
97. *Ibid.*
98. *Ibid.*
99. TNA FD 23/1949, Letter from Mrs PA Lee, DHSS to Mrs DM White, Department of Education and Science, 12 December 1975.
100. TNA FD 23/1949, Note for the file by M. Ashley-Miller, 18 May 1973.
101. Interview with Woodcock.
102. *Ibid.*, p. 93.
103. DL 14039, *The First SCODA Report, 1974* (London: SCODA, 1974); MRC MSS.171/5/21, Proposed Standing Conference on Drug Abuse (SCODA) January 1971.
104. DL 14039, *The First SCODA Report, 1974.*

105. MRC MSS.171/3/18/8, Letter from Bob Searchfield, Coordinator of SCODA to Don Aitken at Release, 5 February 1974.
106. SCODA, *Drug Problems: Where to Get Help* (London: SCODA, 1986).
107. See, for example, DL 26430, Report of the conference 'Ten years after', held 1 July 1976.
108. DL 14039, *The First SCODA Report, 1974*.
109. *Ibid.*
110. Interview conducted by Alex Mold with a SCODA worker 25 February 2005.
111. Joseph Rowntree Trust, *The Future of Voluntary Organisations*, pp. 100–45.
112. DL 32718, *SCODA Annual Report, 1978–79*.
113. TNA MH 154/1192, Background notes on SCODA by DHSS, no date [1976].
114. DL 29050, *SCODA Annual Report, 1976–77*; TNA MH 154/1192, Future Directions for SCODA: a report to the DHSS.
115. TNA MH 154/1192 Memorandum from Mr Ralph to Mr Bruce, Miss McTrusty and Dr Sippert regarding SCODA grant application, 28 March 1977.
116. Interview with SCODA worker.
117. *Ibid.*
118. *Ibid.*
119. DL 23221, *SCODA Annual Report, 1974–75*.
120. Interview with SCODA worker.
121. *Ibid.*
122. *Ibid.*; Interview with Woodcock.
123. *Ibid.*
124. DL 26430, Report of the conference 'Ten years after', 1 July 1976.
125. A.H. Ghodse, 'Conversation with Hamid Ghodse', *Addiction*, 102, (2007), pp. 197–205, p. 200; A.H. Ghodse, 'Casualty departments and the monitoring of drug dependence', *British Medical Journal*, 1, (28 May 1977), pp. 1381–82.
126. Ghodse quoted in A. Jamieson, A. Glanz and S. MacGregor, *Dealing With Drug Misuse: Crisis Intervention in the City* (London: Tavistock, 1984), p. 19.
127. TNA MH 154/1251, Letter from Arthur Blenkinsop, MP [and also SCODA Chairman] to Roland Moyle, Secretary of State for Health, 9 February 1977.
128. See TNA 154/1254, SCODA crisis unit, operational policy documents, 1971–1978.
129. TNA MH 154/1252, Memorandum from Mrs Pearson to Mr Moyes and Mrs Angoy, no date [August 1977?].
130. TNA 154/1254, Operational policy document: City Roads (Crisis Intervention) Ltd., February 1978.
131. Jamieson, Glanz and MacGregor, *Dealing with Drug Misuse*, pp. 38–39. Interview conducted by Alex Mold with Jane MacGregor (former nurse at City Roads) 22 March 2007.
132. Interview conducted by Virginia Berridge with senior social science researcher, 10 February 2005.
133. Jamieson, Glanz and MacGregor, *Dealing with Drug Misuse*, p. 22.
134. Brenton, *The Voluntary Sector in British Social Services*, p. 89.

## 4 Rolling Back the State? The Central Funding Initiative for Drug Services

1. *Hansard Journal of Parliamentary Debates: House of Commons*, 33, (1982–83), p. 704, col. 212.
2. DL 42638, *SCODA Annual Report, 1983–84*, p. 4.
3. MacGregor et al., *Drug Services in England and the Impact of the Central Funding Initiative*.
4. Department of Health Archive, Nelson, Lancashire (hereafter DOHA) OCG/1/1/3/VA/SSI/535, Letter from DHSS to all Regional Health Authorities regarding Treatment and Rehabilitation report of the Advisory Council on the Misuse of Drugs (ACMD); Central Funding Initiative, (HN (83) 13 LASSAL (83) 1), 25 April 1983.
5. *Hansard Journal of Parliamentary Debates: House of Commons*, 33, (1982–83), p. 704, col. 212.
6. Home Office, *Statistics of Drug Addicts Notified to the Home Office, 1988* (London: HMSO, 1989).
7. ACMD, *Treatment and Rehabilitation*, p. 25; Ministry of Health, *Second Report of the Interdepartmental Committee*, p. 8.
8. Home Office, *Tackling Drug Misuse: A Summary of the Government's Strategy* (London: HMSO, 1985), Foreword.
9. V. Berridge, 'The "British system" and its history: myth and reality', in J. Strang and M. Gossop (eds), *Heroin Addiction and the British System. Volume 1 Origins and Evolution* (London: Routledge, 2005), pp. 7–16.
10. S. MacGregor and B. Pimlott, 'Action and inaction in the cities' in S. MacGregor and B. Pimlott (eds), *Tackling the Inner Cities: The 1980s Reviewed, Prospects for the 1990s* (Oxford: Clarendon Press, 1990), p. 9. Although of course state funding of the voluntary sector was nothing new as explored in the previous chapter. For an exploration of a slightly different relationship in a similar period see V. Berridge, 'New social movement or government funded voluntary sector? ASH, (Action on Smoking and Health) science and anti-tobacco activism in the 1970s' in M. Pelling and S. Mandelbrote (eds), *The Practice of Reform in Health, Medicine and Science, 1500–2000. Essays for Charles Webster* (London: Ashgate, 2005), pp. 333–48.
11. Spear, *Heroin Addiction Care and Control*, pp. 41–42.
12. Home Office, *Statistics of Drug Addicts Notified to the Home Office, United Kingdom, 1988*.
13. *Ibid.*
14. G.V. Stimson, 'British drug policies in the 1980s: a preliminary analysis and suggestions for research' in V. Berridge (ed.), *Drugs Research and Policy in Britain: A Review of the 1980s* (Aldershot: Avebury, 1990), pp. 260–81 and J. Mott, 'Notification and the Home Office' in J. Strang and M. Gossop (eds), *Heroin Addiction and Drug Policy: The British System* (Oxford: Oxford University Press, 1994), pp. 271–91, p. 287.
15. S. MacGregor, 'The public debate in the 1980s' in S. MacGregor (ed.), *Drugs and British Society: Responses to a Social Problem in the 1980s* (London: Routledge, 1989), pp. 1–19, p. 3.
16. *The Hansard Journal of Parliamentary Debates: House of Lords*, 30 October 1979, p. 355.

17. For breakdown in regional notifications see ACMD, *Treatment and Rehabilitation*, pp. 121–27.
18. H. Parker, R. Newcombe and K. Bakx, 'The new heroin users: prevalence and characteristics in Wirral, Merseyside', *British Journal of Addiction* (1987), 82, pp. 147–57, p. 147.
19. R. Davenport-Hines, *The Pursuit of Oblivion: A Global History of Narcotics, 1500–2000* (London: Weinfield & Nicholson, 2001), p. 364; R. Power, 'Drug trends since 1968' in Strang and Gossop(eds), *Heroin Addiction and Drug Policy*, pp. 27–41, pp. 34–35.
20. ACMD, *Treatment and Rehabilitation*, p. 130.
21. P. Griffiths, M. Gossop and J. Strang, 'Chasing the dragon: the development of heroin smoking in the United Kingdom' in Strang and Gossop (eds), *Heroin Addiction and Drug Policy*, pp. 121–33, p. 124.
22. G. Stimson, 'The war on heroin: British policy and the international trade in illicit drugs' in N. Dorn and N. South (eds), *A Land Fit For Heroin? Drug Policies, Prevention and Practice* (Basingstoke: Macmillan, 1987), pp. 35–61, pp. 39–41; R. Lewis, 'Flexible hierarchies and dynamic disorder – the trading and distribution of illicit heroin in Britain and Europe, 1970–1990' in Strang and Gossop (eds), *Heroin Addiction and Drug Policy*, pp. 42–65; Spear, *Heroin Addiction Care and Control*, pp. 255–74.
23. MacGregor, 'The public debate in the 1980s', p. 3. The key contemporary paper detailing the link with unemployment was D.F. Peck and M.A. Plant,' Unemployment and illegal drug use: concordant evidence from a prospective study and national trends', *British Medical Journal*, 293, (11 October 1986), pp. 929–32.
24. *Hansard Journal of Political Debates: House of Commons*, 91, (1985–86), p. 296, col. 568.
25. Lowe, *The Welfare State in Britain*, p. 325; J. Harris, 'Tradition and transformation: society and civil society in Britain, 1945–2000' in K. Burk (ed.), *The British Isles Since 1945* (Oxford: Oxford University Press, 2003), pp. 91–125, p. 112.
26. For an overview of the debate on deprivation and drug use see G. Pearson and M. Gilman, 'Drug epidemics in space and time: local diversity, subcultures and social exclusion' in Strang and Gossop(eds), *Heroin Addiction and the British System: Vol 1*, pp. 109–14.
27. M. Kohn, *Narcomania: On Heroin* (London: Faber & Faber, 1987), p. 114.
28. MacGregor et al., *Drug Services in England and the Impact of the Central Funding Initiative*, p. 6, p. 28.
29. Interview conducted by Alex Mold and Virginia Berridge with a Sheffield based psychiatrist, 2 May 2006.
30. A. Burr, 'Increased sale of opiates on the blackmarket [sic.] in the Piccadilly area', *British Medical Journal*, (24 September 1983), pp. 883. Jayne Love and Michael Gossop in their study of the workings of one London DDU between June and December 1983 found that waiting periods of approximately one month between each stage of treatment was not uncommon. See J. Love and M. Gossop, 'The processes of referral and disposal within a London Drug Dependence Clinic', *British Journal of Addiction*, 80, (1985), pp. 435–40, p. 438.
31. Mold, *Heroin*, pp. 55–61.

32. Mold, 'The "British System" and the opening of the Drug Dependence Units', p. 506, p. 509.
33. ACMD, *Treatment and Rehabilitation*, p. 120.
34. 'Drug addiction: British System failing', *Lancet*, (9 January 1982), pp. 83–84, p. 83.
35. S. Mars, 'Peer pressure and imposed consensus: the making of the 1984 Guidelines of Good Clinical Practice in the Treatment of Drug Misuse' in V. Berridge (ed.), *Making health Policy. Networks in research and policy after 1945* (Amsterdam: Rodopi, 2005), pp. 149–82. See also Mold, *Heroin*, Chapters 3 and 4.
36. V. Berridge, 'AIDS and British drug policy: continuity or change?' in V. Berridge and P. Strong (eds), *AIDS and Contemporary History* (Cambridge: Cambridge University Press, 1993), pp. 135–56, p. 141.
37. DL 30658, *SCODA Annual Report, 1977–78*, pp. 2–3.
38. DL 32718, *SCODA Annual Report, 1978–79*, p. 12.
39. DL 36959, *SCODA Annual Report, 1980–81*, p. 6.
40. ACMD, *Treatment and Rehabilitation*, p. 3. For a list of membership see *Ibid.*, pp. 87–88.
41. TNA MH 154/1149, Letter to Minister for Health from Sir Robert Bradshaw, chairman of the ACMD, 25 June 1977.
42. TNA MH 154/1148, Minute from MEG Fogden to Mrs Pearson, 28 February 1977.
43. TNA MH 154/1151, Minutes of the 27th meeting of the ACMD Treatment and Rehabilitation Working Group, 28 June 1979.
44. *Ibid.*
45. ACMD, *Treatment and Rehabilitation*, p. 79.
46. Interview conducted by Virginia Berridge with key social science researcher, 10 February 2005.
47. Interview conducted by Alex Mold and Virginia Berridge with Dr Dorothy Black, 2 May 2006.
48. Interview conducted by Alex Mold and Virginia Berridge with a senior civil servant, 2 May 2006.
49. Berridge, 'AIDS and British drug policy', p. 141; Interview conducted by Alex Mold and Virginia Berridge with Professor Gerry Stimson, 17 May 2006.
50. *Hansard Journal of Political Debates: House of Commons*, 46, (1983–84), pp. 528–31.
51. *Hansard Journal of Political Debates: House of Commons*, 92, (1985–86), p. 937, col. 601.
52. *Ibid.*, col. 602.
53. Interview conducted by authors with Dr Dorothy Black, 2 May 2006.
54. MacGregor and Ettorre, 'From treatment to rehabilitation', p. 145.
55. DOHA OCG/1/1/3/VA/SSI/535, Letter from DHSS to all Regional Health Authorities regarding Treatment and Rehabilitation report of the Advisory Council on the Misuse of Drugs (ACMD); Central Funding Initiative, (HN (83) 13 LASSAL (83) 1), 25 April 1983, p. ii.
56. Interview conducted by authors with Dr Dorothy Black, 2 May 2006.
57. DOHA, DAC/0026/V001, Draft criteria for accepting or rejecting applications, Central Funding Initiative (Drugs) 21 September 1983.
58. Interview between authors and senior civil servant.

59. DOHA, DAC/0014/V0004, Letter from Norman Fowler, Secretary of State for Social Services to Leon Brittan, Home Secretary regarding tackling drug misuse, 30 November 1983.

60. DOHA, DAC/0026/V001, Memorandum from E. Hillier to Mrs Pearson and Mr Ison regarding expenditure on new initiatives, 1986/7 [no date].

61. *Hansard Journal of Parliamentary Debates: House of Commons*, 41, (1982–83), p. 806, col. 397.

62. House of Commons Social Services Committee report, DHSS evidence, 13 March 1985, p. 171.

63. MacGregor et al., *Drug Services in England and the Impact of the Central Funding Initiative*; House of Commons Social Services Committee, *Misuse of Drugs With Special Reference to the Treatment and Rehabilitation of Misusers of Hard Drugs, Session 1984–1985* (London: HMSO, 1984–86), DHSS evidence, 13 March 1985, p. 174.

64. MacGregor et al., *Drug Services in England and the Impact of the Central Funding Initiative*, p. 70.

65. *Ibid.*, p. 28.

66. To some extent DDUs did combine social and medical approaches by aiming to treat and control the drug problem. See Mold, 'The "British System" and the opening of the Drug Dependence Units'.

67. See, for example, Stimson, 'British drug policies in the 1980s'.

68. For an analysis of this see *Ibid.* For the reports see Ministry of Health, *Drug Addiction*, and ACMD, *Treatment and Rehabilitation*.

69. Interview conducted by authors with senior civil servant.

70. For a case study of the different approaches adopted at this time by some regional psychiatrists see J. Strang, 'A model service: turning the generalist on to drugs', in S. MacGregor (ed.), *Drugs and British Society*, pp. 143–69.

71. MacGregor et al., *Drug Services in England and the Impact of the Central Funding Initiative*, p. 45.

72. *Ibid.*, p. 8.

73. ACMD, *Treatment and Rehabilitation*, p. 77.

74. DOHA OCG/1/1/3/VA/SSI/535, Letter from DHSS to all Regional Health Authorities regarding Treatment and Rehabilitation report of the Advisory Council on the Misuse of Drugs (ACMD); Central Funding Initiative, (HN (83) 13 LASSAL (83) 1), 25 April 1983, p. 1.

75. DHSS evidence of House of Commons Social Services Committee, p. 170.

76. Interview conducted by authors with senior civil servant.

77. *Ibid.*

78. MacGregor et al., *Drug Services in England and the Impact of the Central Funding Initiative*, pp. 71–74.

79. House of Commons Social Services Committee, *The Misuse of Drugs*, p. liii, p. xlvii.

80. Interview between authors and senior social science researcher.

81. DOHA, JR/01980565/V0001A, Paper for discussion at drugs client team meeting on research project on CFI, 31 October 1985.

82. DOHA, JR/01980565/V0001A, Letter from Anne Kauder, Office of the Chief Scientist, DHSS to Susanne MacGregor, 17 January 1986.

83. DOHA, JR/01980565/V0001A, Paper for discussion at drugs client team meeting on research project on CFI, 31 October 1985.

84. N. Black, 'The NHS research and development programme: the first five years, 1991–6', seminar paper at London School of Hygiene and Tropical Medicine, 21 January 1997.
85. MacGregor et al., *Drug Services in England and the Impact of the Central Funding Initiative*, p. 6.
86. *Ibid.*, p. 13.
87. DOHA JR/0198/0401, Eric Blakebrough, 'Choosing the wrong vein', *The Guardian* 6 April 1983. Kaleidoscope latter received two grants under the CFI. See MacGregor et al., *Services in England and the Impact of the Central Funding Initiative*, p. 73.
88. MacGregor et al., *Services in England and the Impact of the Central Funding Initiative*, p. 7.
89. DOHA DAC/0026/V001, Memo from Hillier to Nye regarding Drug Initiative Evaluation, 5 September 1983.
90. Interview with Yates.
91. MacGregor et al., *Services in England and the Impact of the Central Funding Initiative*, p. 45.
92. Interview conducted by Virginia Berridge with Roger Howard, 20 March 2006.
93. On urban initiatives see MacGregor and Pimlott, 'Action and inaction in the cities'.
94. DOHA, DAC/0007/V0004, Note on Central Initiatives by John H James, 30 April 1986; DOHA, DAC/0026/V001, Memorandum from DC Nye to Mr Alderman, Miss Davies, Mr Hillier, Mr Lutterloch, Mr Pagan and Mr Woolley, regarding new initiatives, 14 December 1983.
95. DOHA, DAC/0007/V0004, Note on Central Initiatives by John H James, 30 April 1986.
96. On changes in the welfare state in this period see Lowe, *The Welfare State*, pp. 317–27.
97. *Ibid.*, pp. 325–26.
98. M. Harris, C. Rochester and P. Halfpenny, 'Voluntary organisations and social policy: twenty years of change' in M. Harris and C. Rochester (eds), *Voluntary Organisations and Social Policy in Britain: Perspectives on Change and Choice* (Basingstoke: Palgrave, 2001), p. 3.
99. Lewis, 'Developing the mixed economy', pp. 173–92.
100. Kendall and Knapp, *The Voluntary Sector in the UK*, p. 138; Deakin, 'The perils of partnership', p. 54.
101. Lowe, *The Welfare State in Britain*, p. 320.
102. Lewis, 'Developing the mixed economy', pp. 183–91.
103. MacGregor and Pimlott, 'Action and inaction', p. 9.
104. DOHA, DAC/0014/V0004, Draft submission reporting on the consultation carried out in the wake of the Treatment and Rehabilitation report, November 1983.
105. DOHA, DAC/0014/V0004, Draft letter to the Home Secretary, November 1983.
106. Lowe, *The Welfare State in Britain*, pp. 325–26, p. 333.
107. V. Berridge, *AIDS in the UK: The Making of Policy, 1981–1994* (Oxford: Oxford University Press, 1996), p. 222.
108. ACMD, *AIDS and Drug Misuse, Part One* (London: HMSO, 1988).

## 5   Activism and Health: The Impact of AIDS

1. N. Deakin, *In Search of Civil Society*, p. 52.
2. Billis, *Organising Public and Voluntary Agencies*. See also V. Berridge, ' "Unambiguous voluntarism?" AIDS and the voluntary sector in the United Kingdom, 1981–1992' in C. Hannaway, V.A. Harden and J. Parascandola (eds), *AIDS and the Public debate* (Amsterdam: IOS Press, 1995), pp. 153–69.
3. See for example, M. Stacey, 'The health service consumer: a sociological mis-conception', *The Sociological Review Monograph: The Sociology of the National Health Service*, 22, (1978), pp. 194–200 and C.J. Ham, 'Power, patients and pluralism' in K. Barnard and K. Lee (eds), *Conflicts in the National Health Service* (London: Croom Helm, 1977), pp. 99–110.
4. Webster, *The National Health Service: A Political History*, pp. 187–200.
5. C. Hogg, *Patients, Power and Politics: From Patients to Citizens* (London: Sage, 1999), pp. 42–49.
6. Cm 1599, *The Citizen's Charter: Raising the Standard* (London: HMSO, 1991).
7. Cm 4818-I, *NHS Plan, 2000: A Plan for Investment, A Plan for Reform* (London: TSO, 2000).
8. See also L. Robinson, *Gay Men and the Left: How the Personal Got Political* (Manchester: Manchester University Press, 2007).
9. Interview with Janet Green by Janet Foster, AIDS Social History Programme, 28 January 1993.
10. Interview by Virginia Berridge with Jonathan Grimshaw, 2 March, 1990.
11. Interview by Virginia Berridge with Tony Coxon, 15 March, 1990.
12. S. Hagard, speech to National Conference of Voluntary AIDS Helplines, 1987.
13. As discussed in V. Berridge, *AIDS in the UK: The Making of Policy, 1981–1994* (Oxford: Oxford University Press, 1996), especially pp. 268–73.
14. Discussed below and in the Conclusion.
15. Interview by Virginia Berridge with AIDS voluntary sector worker, January 1989.
16. Interview by Virginia Berridge with AIDS voluntary sector worker, June 1993.
17. See in Chapter 6.
18. See Berridge, *AIDS in the UK*, p. 91.
19. Interview with Bill Nelles by Virginia Berridge, 1 February, 1990.
20. Interview with Bill Nelles by Virginia Berridge, 1 February 1990, also inter-view with Tony Whitehead by Virginia Berridge and Phil Strong, 6 July 1989.
21. Interview with Bill Nelles by Virginia Berridge, 1 February 1990. Other interviews also allude to this rule. There was a SCODA recommendation that ex-drug users should be drug-free for at least 2 years before working in a voluntary drug service. This came to be interpreted as an official rule, but was really just an unofficial guideline.
22. J.R. Robertson, A.B.V. Bucknall, P.D. Welsby et al., 'Epidemic of AIDS related virus (HTLVIII/LAV) infection among intravenous drug users', *British Medical Journal*, 292, (1986), p. 527; see also Berridge, *AIDS in the UK*, pp. 92–3.
23. Interview with Janet Green, 1993.
24. Interview with Betsy Ettorre by Virginia Berridge, 1 February 1990.
25. Interview by Alex Mold with a user activist 9 July 2005.

26. Scottish Home and Health Department, *HIV Infection in Scotland: the Report of the Scottish Committee on HIV Infection and Intravenous Drug Misuse* (Edinburgh: Scottish Home and Health Department, 1986)
27. Advisory Council on the Misuse of Drugs, *AIDS and Drug Misuse: Part 1* (London: HMSO, 1988)
28. See Mold, *Heroin.*
29. V. Berridge, D. Christie and E.M. Tansey (eds.), *Public Health in the 1980s and 90s: Decline and Rise?* (London: Wellcome Centre, 2006).
30. Interview conducted by Alex Mold with a drug user activist, 27 July 2006.
31. DL 48602, *SCODA Annual Report, 1985–86*, p. 10.
32. Interview by Alex Mold with a drug user activist, 27 July 2006.
33. Contemporary Medical Archives Centre, Wellcome Library, London (hereafter CMAC) Private Papers of Dr Ann Dally (hereafter PP/DAL) B/4/1/3/1, DIG Meeting 29 April 1987.
34. CMAC PP/DAL/B/4/1/3/2, Letter from DIG to John Calderan at Theodore Goddard, 17 July 1987.
35. Berridge AIDS book has more on this.
36. Interview by Alex Mold with a user activist, 13 December 2005.
37. *Ibid.*
38. Virginia Berridge's notes of the Hatfield conference.
39. *SCODA Newsletter*, October/November, 1991 'Users have their say'; P. McDermott, 'User run, user friendly: why drug workers should empower drug users into their jobs', *Druglink* (November/December, 1992), p. 11.
40. D. Tops, 'Stretching the limits of drug policies: an uneasy balancing act', in Anker, et al. (eds), *Drug Users and Spaces for Legitimate Action*, pp. 61–83, pp. 65–66; A. Efthimiou-Mordaunt, 'Junkies in the House of the Lord', unpublished MSc dissertation, University of London, 2004, pp. 9–13.
41. McDermott, 'User run, user friendly', p. 11; Jorgen Anker, 'Active drug users – struggling for rights and recognition', pp. 37–41; N. Crofts, 'A history of peer-based drug user groups in Australia', *Journal of Drug Issues*, 25, (1993), pp. 599–616.
42. S.R.. Friedman, D.C. Des Jarlais et al., 'AIDS and self organisation among intravenous drug users', *International Journal of the Addictions*, 22:3, (1987), pp. 210–19; S.R..Friedman and C. Casriel, 'Drug users' organisations and AIDS Policy', *AIDS and Public Policy Journal*, 3:2, (1988), pp. 30–36; S.R..Friedman, W. M.de Jong and D.C.des Jarlais, 'Problems and dynamics of organising intravenous drugs users for AIDS prevention', *Health Education Research,* 3:1, (1988), pp. 49–57. See also interviews with UK user activists who mention his influence.
43. Interview with a user activist, 13 December 2005.
44. Berridge, *AIDS in the UK*, p. 18.
45. *Ibid.*, pp. 275–76; S. Epstein, *Impure Science. AIDS, Activism and the Politics of Knowledge* (Berkeley: University of California Press, 1996).
46. V. Berridge, 'The "British system" and its history: myth and reality' in J. Strang and M. Gossop (eds), *Heroin Addiction and the British system volume 1 Origins and evolution* (London: Routledge, 2005), pp. 7–16.
47. Interview by Alex Mold with SCODA worker, 25 February 2005.
48. This is discussed in the following chapter.
49. Interview by Virginia Berridge with DH civil servant May 1993.
50. J. Polkinghorne, M. Gossop and J. Strang, 'The Government Task Force and its review of drug treatment services. The promotion of an evidence based

approach' in J. Strang and M. Gossop (eds), *Heroin Addiction and the British System Vol 2*, pp. 198–205.

51. Interview by Virginia Berridge with a GP working in the drugs field, 10 December 2006.
52. See, for example, Royal College of General Practitioners, *Guidance for the use of Methadone for the Treatment of Opioid Dependence in Primary Care* (London: RCGP, 2005).This had the RCGP Substance Misuse Unit, the RCGP Sex Drugs and HIV Task Group and The Alliance on its mast head.
53. Interview with a GP working in the drugs field.
54. Interview by Virginia Berridge and Alex Mold with Bill Nelles, 10 March 2006.
55. Department of Health, *Drug Misuse and Dependence: Guidelines on Clinical Management – Update June 2007, Consultation Draft* (London: Department of Health, 2007), p. 91; Royal College of General Practitioners, *Guidance for the use of Buprenorphine for the Treatment of Opioid Dependence in Primary Care* (London: RCGP, 2004), p. 7.
56. Southwell, *Guide to Involving and Empowering Users*.
57. Interview with a user activist, 23 August 2005.
58. Interview with Nelles.
59. *Ibid.*
60. Interview with a user activist, 23 August 2005.
61. D. Best and A. Campbell, *Summary of the NTA's National Prescribing Audit* (NTA: London, 2006), p. 5.
62. Interview by Virginia Berridge and Alex Mold with therapeutic community worker, 10 March 2006.
63. Interview by Virginia Berridge with NTA worker, 30 March 2007.
64. See Conclusion.
65. T. Whitehead, 'The voluntary sector: five years on' in E. Carter and S. Watney (eds), *Taking Liberties: AIDS and Cultural Politics* (London: Serpent's Tail, 1989), pp. 107–11.

## 6 Business Models or the Revival of the State?

1. Kendall and Knapp, *The Voluntary Sector in the UK*, p. 206.
2. Lowe, *The Welfare State in Britain Since 1945*, p. 352.
3. Lewis, 'Developing the mixed economy'.
4. Kendall and Knapp, *The Voluntary Sector in the UK*, p. 9.
5. Interview with senior social policy researcher, by Virginia Berridge, 10 February 2005.
6. Interview with Roger Howard by Virginia Berridge, 20 March 2006.
7. J. Kendall, 'The mainstreaming of the third sector into public policy in England in the late 1990s: whys and wherefores', Civil Society Working Paper 2, January 2000, http://www.lse.ac.uk/collections/CCS/pdf/CSWP/cswp2.pdf, accessed 28 July 2009.
8. Lowe, *The Welfare State in Britain Since 1945*, p. 350.
9. Roy Griffiths, *Community Care: An Agenda For Action: A Report to the Secretary of State for Social Services* (London: HMSO, 1988).

10. Webster, *The National Health Service*, p. 196.
11. Lowe, *The Welfare State in Britain Since 1945*, p. 351.
12. Lewis, 'Mixed economy of welfare', pp. 186–87.
13. Kendall and Knapp, *The Voluntary Sector in the UK*, pp. 230–31.
14. Lewis, 'Mixed economy of welfare', p. 179.
15. Kendall and Knapp, *The Voluntary Sector in the UK*, p. 233.
16. Home Office, *Efficiency Scrutiny of Government Funding of the Voluntary Sector: Profiting from Partnership* (London: Home Office, 1990).
17. Lewis, 'The relationship between the voluntary sector and the state in Britain in the 1990s', p. 261.
18. B. Knight, *Voluntary Action* (London: Home Office, 1993).
19. The report still provokes controversy, see B. Knight, 'Should charities deliver public services? No says Barry Knight', *The Guardian*, 1 October 2003, http://www.guardian.co.uk/society/2003/oct/01/thinktanks.futureforpublicservices, accessed 20 November 2007.
20. Lewis, 'The relationship between the voluntary sector and the state in Britain in the 1990s', p. 263. See also P. 6 and D. Leat, 'Inventing the British voluntary sector by committee: from Wolfenden to Deakin', *Non Profit Studies*, 1, (1996), pp. 33–45.
21. Lewis, 'The relationship between the voluntary sector and the state in Britain in the 1990s', pp. 263–64.
22. Cm 4100, *Compact on Relations Between Government and the Voluntary and Community Sector in England* (London: HMSO, 1998).
23. Lewis, 'The relationship between the voluntary sector and the state in Britain in the 1990s', p. 264.
24. See, for example, HM Treasury, *Exploring the Role of the Third Sector in Public Service Delivery and Reform: A Discussion Document* (London: HM Treasury, 2004).
25. Interview with ISDD worker.
26. See http://www.cabinetoffice.gov.uk/third_sector/about_us.aspx, accessed 7 December 2007.
27. NCVO, *The UK Voluntary Sector Almanac 2006*, p. 8.
28. These themes are discussed in greater detail in the Conclusion.
29. Home Office, *Tackling Drug Misuse: A Summary of the Government's Strategy* (London: HMSO, 1985).
30. Cmd 4221, *The Government's Expenditure Plans 1990–00 to 2001–02* (London: The Stationary Office, 1999).
31. British Medical Association, *The Misuse of Drugs* (London: Harwood Academic Publishers, 1997), p. 27.
32. Cmd 2846, *Tackling Drugs Together: A Strategy for England 1995–1998* (London: The Stationary Office, 1995).
33. Cmd 3945, *Tackling Drugs to Build a Better Britain* (London: The Stationary Office, 1998).
34. G. Stimson, 'Blair declares war the unhealthy state of British drug policy', *International Journal of Drug Policy*, 11, (2000), p. 260.
35. See http://www.nta.nhs.uk/, accessed 7 December 2007.
36. K. Duke, 'Out of crime and into treatment? The criminalization of contemporary drug policy since *Tackling Drugs Together*,' *Drugs: Education, Prevention and Policy*, 13:5, (2006), p. 409.

37. M. Hough, 'Balancing public health and criminal justice interventions,' *International Journal of Drug Policy*, 12, (2001), p. 429.
38. M. Gossop, *Drug Misuse Treatment and Reductions in Crime: Findings From the National Treatment Outcome Research Study (NTORS)* (London: NTA, 2005), p. 3.
39. *Tackling Drugs to Build a Better Britain*, p. 1.
40. G. Kothari, J. Marsden and J. Strang, 'Opportunities and obstacles for effective treatment of drug misusers in the criminal justice system in England and Wales,' *British Journal of Criminology*, 42, (2002), p. 412.
41. Stimson, 'Blair declares war', pp. 259–60.
42. Berridge, 'AIDS in British drug policy'.
43. Gossop, *Drug Misuse and Reductions in Crime*, p. 4.
44. E. Finch and M. Ashton, 'Treatment to order: the new drug treatment and testing orders' in Strang and Gossop(eds), *Heroin Addiction and the British System Vol 2*, p. 189.
45. Duke, 'Out of crime and into treatment?' p. 412.
46. Interview conducted by Virginia Berridge with Mike Trace, 24 January 2007.
47. Cited in 'We can help each other', *Guardian*, 22 February 2007.
48. *Ibid.*
49. *Ibid.*
50. Interview with Jane McGregor by Alex Mold, 22 March 2007.
51. Alison Chesney, obituary *Guardian*, 3 August, 2006.
52. Mary McGloin letter to Virginia Berridge 12 September 1994; also The Cranstoun Projects, Annual Report and Accounts, 1992–93 and 1993–94; The Cranstoun Projects, Prisoners' Resource Service, Annual Report, 1993–94.
53. Interview with Gerry Stimson by Alex Mold and Virginia Berridge, 17 May 2006.
54. See Thom, *Dealing With Drink*, pp. 91–92.
55. Turning Point website the history of Turning Point, accessed 15 February 2007.
56. V. Berridge, *Marketing Health. Smoking and the Discourse of Public Health in Britain, 1945–c.2000* (Oxford: Oxford University Press, 2007), pp. 161–84.
57. Timeline, also Arbery's c.v. in VB's possession prior to interview which could not take place because of his stroke.
58. Interview with Roger Howard.
59. Interview with senior manager at Turning Point by Virginia Berridge, 12 March 2007.
60. Interview with Victor Adebowale. Partnership Portal news, www.our partnership.org.uk, accessed 12 March 2007.
61. *Ibid.*
62. Interview with an ADFAM worker by Virginia Berridge, 22 March 2007.
63. Social Enterprise Coalition, *There's More to Business Than you Think. A Guide to Social Enterprise.* (2003)
64. 'Our background' from Exchange supplies website, accessed 26 February 2007.
65. Interview with Andrew Preston by Virginia Berridge, 27 March 2007.

66. See discussion in Chapter 5.
67. S. Clement and J. Strang, 'The rise and fall of the Community Drugs team. The gap between aspiration and achievement' in J. Strang and M. Gossop (eds), *Heroin Addiction and the British System Volume 2* (London: Routledge, 2005), pp. 94–104.
68. 'SCODA members in two minds about admitting statutory agencies', *Druglink*, 9:2, (March/April, 1994), p. 5.
69. DOHA DAC0090/V002, 'Employment of a management specialist to aid in the restructuring of SCODA', memorandum from Margaret Jackman to Ms Hampson, Mr Barratt and Mr Dudley, 31 January 1994.
70. DOHA DAC0090/V002, 'SCODA and Alcohol Concern', memorandum from Dr M Farrell to Ms Jackman, 3 June 1992.
71. DOHA DAC0090/V002 'Draft specification for employment of a consultant to work with SCODA', attached to 'Employment of a management specialist to aid in the restructuring of SCODA', memorandum from Margaret Jackman to Ms Hampson, Mr Barratt and Mr Dudley, 31 January 1994.
72. M. Ashton, 'Radical change risks crisis of confidence in SCODA', *Druglink*, 10:1, (Jan/Feb 1995), p. 5.
73. *Ibid*, p. 8.
74. No mention of the report was made in the SCODA annual report or its regular newsletters.
75. Ashton, 'Radical change'.
76. Interview by Virginia Berridge with NTA worker, 13 March 2007.
77. SCODA, *Getting Drug Users Involved, Good Practice in Local Treatment and Planning* (London: SCODA, 1997).
78. P. Farley, Can these bones live?' *Druglink*, 14:1, (Jan/Feb1999), pp. 18–19.
79. Virginia Berridge papers.Chairman's report for the meeting of council on 23 September 1998.
80. M. Barnes, 'Back to the roots', *Druglink*, (Jan/Feb 2007), pp. 8–9, discussed some of the difficulties.
81. Home Office and Department of Health, *Review of Second Tier Activity in the Drugs Sector: Final Report* (Cordis Bright, December 2006).
82. Sandra M. Nutley, Isabel Walter and Huw T.O. Davies, *Using Evidence. How Research Can Inform Public Services* (Bristol: The Policy Press, 2007), p. 250; S. M. Nutley, N. Bland and I. C. Walter, 'The Institutional arrangements for connecting evidence and policy: the case of drug misuse', *Public Policy and Administration*, 17:3, (2002), pp. 76–94.
83. Annette Dale Perera, 'National treatment agency. Building castles on sand?' *Druglink*, (May/June 2001), pp. 19–21.
84. Interview with NTA worker.
85. Interview with drug policy official.
86. Interview by Virginia Berridge with Sally Taylorson, 13 April, 2007.
87. See below Chapter 7.
88. Interview with Taylorson.
89. Trace's connections are well known. Taylorson's first husband had been a drug user and she had also worked for Release.
90. NTA website, accessed 21 January 2008. Mc Dermott's biography as a Board member made no mention of his user involvement.

# 7   Users: Service Users and the Drug User Movement

1. 'Drug User Activists Declare a Statement About the International Network of People Who Use Drugs', 30 April 2006, http://www.hardcoreharmreducer. be, accessed 28 July 2008.
2. Interview conducted by Alex Mold with a user activist, 6 July 2005.
3. Health and Social Care Act, 2001, http://www.opsi.gov.uk/ACTS/acts2001/20020015.htm, accessed 28 July 2008.
4. A. Mold, 'Patient groups and the construction of the patient as "consumer" in Britain since the 1960s', unpublished seminar paper for the School of Population & Health Sciences Seminar, University of Birmingham, 6 May 2009.
5. See, for example, R. Forster, and J. Gabe, 'Voice or choice? Patient and public involvement in the National Health Service in England under New Labour', *International Journal of Health Services*, 38:2, (2008), pp. 333–56; I. Greener, 'Who choosing what? The evolution of the use of "choice" in the NHS and its importance for New Labour' in C. Bochel, N. Ellison and M. Powell (eds) *Social Policy Review 15: UK and International Perspectives* (Bristol: The Social Policy Press, 2003), pp. 49–68; P. 6, 'Giving consumers of British public services more choice: what can be learned from recent history?', *Journal of Social Policy*, 32:2, (2003), pp. 239–70.
6. See for example, Department of Health, *NHS Plan: A Plan for Investment, A Plan for Reform* (London: The Stationery Office, 2000); Department of Health, *Choosing Health: Making Healthy Choices Easier* (London: The Stationary Office, 2004); Department of Health *Our Health, Our Care, Our Say* (London, Department of Health, 2006); Department of Health, *Choice Matters 2007–08: Putting Patients in Control*, (London: Department of Health, 2007); Department of Health, *NHS Choices: Delivering for the NHS*, (London, Department of Health, 2008); Department of Health, *NHS Constitution* (London: Department of Health, 2009).
7. On some of the problems surrounding choice, see J. Clarke, N. Smith, and E. Vidler, 'The indeterminacy of choice: political, policy and organisational implications', *Social Policy and Society*, 5:3, (2006), pp. 327–36; J. Clarke, ' "It's not like shopping": citizens, consumers and the reform of public services' in M. Bevir and F. Trentmann (eds), *Governance, Consumers and Citizens: Agency and Resistance in Contemporary Politics* (Basingstoke: Palgrave, 2007), pp. 97–118.
8. E. Andersson, S. Creasy and J. Tritter, 'Overview: does patient and public involvement matter?' in E. Andersson, J. Tritter and R. Wilson (eds), *Healthy Democracy: The Future of Involvement in Health and Social Care* (London: Involve and NHS Centre for Involvement, 2007), pp. 7–17.
9. R. Baggott, J. Allsop and K. Jones, *Speaking for Patients and Carers: Health Consumer Groups and the Policy Process* (Basingstoke: Palgrave, 2005), pp. 1–6.
10. Andersson, Creasy and Tritter, 'Overview', pp. 10–11.
11. D. Wanless, *Securing our Future Health: Taking a Long Term View* (London: HM Treasury, 2002).
12. Cm 6737, *Our Health, Our Care, Our Say: A New Direction for Community Health Services* (London: TSO, 2006).
13. F. Branfield and P. Beresford, *Making User Involvement Work: Supporting Service User Networking and Knowledge* (York: Joseph Rowntree Foundation, 2006), p. vii.

14. Interview with a drug user activist, 30 May 2006.
15. NTA, *NTA Guidance For Local Partnerships on User and Carer Involvement* (London: NTA, 2006).
16. National Treatment Agency, *Being Heard: Notable Examples of User and Carer Organisations* (NTA: London, 2004), p. 2.
17. See http://www.nta.nhs.uk/publications/openingdoors1.htm, accessed 17 May 2005. For more on the expert patient scheme see Department of Health, *The Expert Patient: A New Approach to Chronic Disease Management for the 21st Century* (London: TSO, 2001).
18. See http://www.nta.nhs.uk/areas/users_and_carers/toolkit.aspx, accessed 28 July 2008.
19. NTA, *The NTA's First Annual User Satisfaction Survey 2005* (London: NTA, 2006).
20. NTA, *NTA Guidance For Local Partnerships on User and Carer Involvement*.
21. Interview with Bill Nelles; Interview with Mike Trace.
22. NTA, *NTA Guidance For Local Partnerships on User and Carer Involvement*.
23. *Ibid.*; Anon, 'Learning from the experts: setting up a user group', *Drink and Drug News*, 18 April 2005, pp. 10–11; Interview with user activist, 30 May 2006.
24. Interview with a drug user activist, 27 July 2006; M. Southwell, *A Guide to Involving and Empowering Drug Users*, no date [2003?], http://www.harm reductionnetwork.mb.ca/nta.pdf, accessed 23 May 2007.
25. NTA, *NTA Guidance For Local Partnerships on User and Carer Involvement*.
26. NTA, *Being Heard: Notable Examples of User and Carer Organisations* (NTA: London, 2004).
27. *Ibid.*; G. Daniels, 'The real experts perspective', http://www.nta.nhs.uk/events/pdfs/26_The_real_experts_perspective_speech.pdf, accessed 29 July 2008; S. Moralioglu, 'Learning from experience', *The Guardian*, 7 March 2007.
28. M. Daly, 'More questions than answers', *Druglink*, (May/June 2005), pp. 14–15.
29. Moralioglu, 'Learning from experience'.
30. Daniels, 'The real experts perspective'; Interview with a drug user activist, 27 July 2006.
31. Daniels, 'The real experts perspective'.
32. M. Southwell and T. Miller, 'Reflecting on the history and lessons of attempts to form a national drug users movement' downloaded from http://www.traffasi.com/modules.php?op=modload&name=PagEd&file=index&print..., accessed 31 August 2005.
33. Southwell, *A Guide to Involving and Empowering Drug Users*; Interview with Nelles.
34. Interview with a drug user activist, 30 May 2006.
35. *Black Poppy*, Issue 4, no date [2000?]; *Black Poppy*, Issue 8, 2003.
36. Interview conducted by Alex Mold with a drug user activist, 19 July 2005.
37. Interview with a drug user activist, 6 July 2005.
38. *The Users Voice*, Issue 4, (December 1998/January 1999).
39. *Ibid.*
40. Interview conducted by Alex Mold with a user activist, 20 July 2006.
41. See, for example, the National Drug Treatment Conference Programme, 2006, http://www.exchangesupplies.org/conferences/2006_NDTC/programme.html, accessed 3 March 2007.
42. Interview with a user activist, 30 May 2005.

43. Interview with a user activist, 25 July 2005.
44. Interview with a user activist, 6 July 2005.
45. Interview with a user activist, 25 July 2005.
46. Interview with a user activist, 13 December 2005.
47. E. O'Mara, 'What's the score? Black Poppy with no strings attached', *Black Poppy*, Issue 4, (no date, 2000).
48. Interview conducted by Alex Mold with a user activist, 13 April 2006.
49. Interview with user activist, 30 May 2006.
50. 'Service Users Charter', Methadone Alliance, no date, version 1.2.
51. Peter McDermott, *Out-patient Treatment for Heroin Addiction: A Service Users' Guide to Rights and Responsibilities* (Manchester: Lifeline Publications, 2003), p. 7.
52. Interview with a user activist, 13 December 2005.
53. Interview with a user activist, 6 July 2005.
54. Interview conducted by Alex Mold with a worker at Transform, 5 July 2006. Although Transform no longer publicly state this as a reason for what they now call 'drug policy reform' rather than 'legalisation', the organisation still believes this to be true.
55. Gary Sutton, 'The flash, the crash and the quest for the cash', *Black Poppy*, Issue 4, (no date, 2000).
56. Interview with a user activist, 30 May 2006.
57. O'Mara, 'What's the score?' *Black Poppy*, Issue 4, (no date, 2000).
58. Interview with user activist, 30 May 2006. On Thailand's clampdown on illegal drug use and Amnesty International's concerns see http://news.bbc.co.uk/1/hi/world/asia-pacific/2793763.stm, accessed 22 June 2007 and http://web.amnesty.org/library/index/engasa390012003, accessed 22 June 2007.
59. Interview conducted by Alex Mold with user activist and writer for *Black Poppy*, 23 August 2005.
60. Gary Sutton, 'Working in the field', *The User's Voice*, Issue 5, (March 1999).
61. See http://www.famanon.org.uk/index.html, accessed 22 January 2007. For a study of drug user family groups in the late 1980s, see also P. Gay, *Getting Together: A Study of Self-Help Groups for Drug Users Families* (London: PSI Publications, 1989).
62. Home Office, *Updated Drug Strategy* (London: HMSO, 2002).
63. NTA, *Expenses Policy for Service Users and Carers* (London: NTA, 2005).
64. NTA, *NTA Guidance for Local Partnerships on User and Carer Involvement* (NTA: London, 2006).
65. Interview with a user activist, 30 May 2006.
66. Interview by Virginia Berridge with carer organisation worker, 22 March 2007.
67. NTA, *Supporting and Involving Carers* (London: NTA, 2006).
68. ACMD, *Hidden Harm: Responding to the Needs of the Children of Problem Drug Users* (London: HMSO, 2003), p. 3.
69. B. Wood, *Patient Power? The Politics of Patients' Associations in Britain and America* (Buckingham: Open University Press, 2000), pp. 187–89.
70. M. Barnes, S. Harrison, M. Mort and P. Shardlow, *Unequal Partners: User Groups and Community Care* (Bristol: Policy Press, 1999), p. 107.
71. B. Salter, 'Patients and doctors: reformulating the UK health policy community', *Social Science and Medicine*, 57, (2003), pp. 927–36, p. 930.

72. Baggott, Allsop and Jones, *Speaking for Patients and Carers*, p. 227.
73. See, for example, G.V. Stimson, ' "Blair declares war": the unhealthy state of British drug policy', *International Journal of Drug Policy*, 11, (2000), pp. 259–64.
74. NTA, *First Annual User Satisfaction Survey*, p. 2, p. 4.
75. EATA, *Service User Views of Treatment: Research Conducted for the Audit Commission* (EATA: London, 2004).
76. J. Fischer et al., *Drug User Involvement in Treatment Decisions* (York: Joseph Rowntree Foundation, 2007). This finding was also echoed by Sue Patterson and her colleagues in a similar study. See S. Patterson, T. Weaver, K. Agath, E. Albert and T. Rhodes, ' "They can't solve the problem without us": a qualitative study of stakeholder perspectives on user involvement in drug treatment services in England', *Health and Social Care in the Community*, 17:1, (2008), pp. 54–62.
77. Fischer et al., *Drug User Involvement in Treatment Decisions*, p. 12.
78. *Ibid.*, p. 37.
79. *Ibid.*, p. 18.
80. *Ibid.*, p. 1.
81. Patterson et al., ' "They can't solve the problem without us," ' p. 58.
82. J. Birchall and R. Simmons, *User Power: The Participation of Users in Public Services* (London: National Consumer Council, 2004), p. 42; Branfield and Beresford, *Making User Involvement Work*, p. vii.
83. Interview with a drug user activist, 30 May 2006.
84. Interview with a drug user activist, 27 July 2006.
85. Interview conducted by Virginia Berridge and Alex Mold with a drug addiction psychiatrist, 10 March 2006.
86. Patterson et al., ' "They can't solve the problem without us," ' p. 60.
87. S. Harrison and M. Mort, 'Which champions, which people? Public and user involvement in health care as a technology of legitimation', *Social Policy and Administration*, 32:1, (1998), pp. 60–70.
88. Barnes et al., *Unequal Partners*, p. 107.
89. Branfield and Beresford, *Making User Involvement Work*, p. 36.
90. Interview with a user activist, 25 July 2005. See also E. O'Mara, 'Opinion: Five years ago there was there was a promise of better things to come through "user involvement". In hindsight have those promises been fulfilled? No.', *Drugs and Alcohol Today*, 4:1, (2004), pp. 18–19, p. 18.
91. Baggott, Allsop and Jones, *Speaking for Patients and Carers*, pp. 228–29.
92. See, for example, P. Beresford and A. Wilson, 'Genes spell danger: mental health service users/survivors, bioethics and control', *Disability and Society*, 17:5, (2002), pp. 541–53; N. Crossley and M. Crossley, 'Patient voices, social movements and habitus. How psychiatric survivors speak out', *Social Science and Medicine*, 52, (2001), pp. 1477–89; A. Rogers, and D. Pilgrim, 'Pulling down churches: accounting for the British mental health users movement', *Sociology of Health and Illness*, 13:2, (1991), pp. 129–48.
93. On the stigmatisation of drug users see S.R. Friedman, 'The political economy of drug user scapegoating – the philosophy and politics of resistance', *Drugs: Education, Prevention and Policy*, 5:1, (1998), pp. 15–32, and Patterson et al., ' "They can't solve the problem without us",' p. 60.

94. S.R. Friedman, 'Theoretical bases for understanding drug users' organisations', *International Journal of Drug Policy*, 7:4, (1996), pp. 212–19, pp. 216–17.
95. Interview with a user activist, 25 July 2005.
96. See for example, K. Duke, 'Out of crime and into treatment? The criminalization of contemporary drug policy since *Tackling Drugs Together*', *Drugs: Education, Prevention and Policy*, 13:5, (2006), pp. 409–15; and Stimson, 'Blair declares war'.
97. E. Finch and M. Ashton, 'Treatment to order: the new drug treatment and testing orders' in Strang and Gossop (eds), *Heroin Addiction and the British System Vol 2*, pp. 187–97.
98. Michelle Cave, 'Fighting for survival', *Drink and Drug News*, 8 May 2006, p. 12.
99. Interview with a user activist, 25 August 2005.
100. *Ibid.*, Southwell and Miller 'Reflecting on the history and lessons of attempts to form a national drug users movement'.
101. *Ibid.*
102. Interview with a user activist.
103. Interview with a drug user activist, 27 July 2006.
104. H. Shapiro, 'Nothing about us, without us: user involvement, past, present and future', *Druglink*, (May/June 2005), pp. 10–11.
105. Berridge, *AIDS in the UK*, pp. 202–5.
106. Branfield and Beresford, *Making User Involvement Work*; Harrison and Mort, 'Which champions, which people?', p. 65.
107. Interview with a SCODA worker, 25 February 2005.
108. Interview with leading drug addiction psychiatrist.
109. Interview with user activists.
110. Interview with a user activist, 30 May 2006.
111. Interview with a user activist, 25 July 2005.
112. Interview with a user activist, 13 December 2005; Interview with a user activist 30 May 2005; Interview conducted by Alex Mold and Virginia Berridge with Gerry Stimson, 17 May 2006; Crofts, 'A history of peer-based drug user groups in Australia'; *Users Voice*; *Black Poppy*.
113. Anker, Asmussen, Kouvonen and Tops, 'Introduction', p. 7.
114. Interview with Stimson.
115. See http://www.inpud.org/, accessed 22 June 2007.
116. Interview with IHRA official.
117. See http://www.encod.org/, accessed 22 June 2007.
118. Interview with Stimson.
119. Interview with Stimson; Interview with Kushlick.
120. Anker et al., 'Introduction', p. 7.
121. Salter, 'Patients and doctors', p. 933.

## Conclusion

1. HM Treasury, *The Future Role of the Third Sector in Social and Economic Regeneration: Final Report* (London: HM Treasury, 2007), p. 3.
2. Lewis, 'New Labour's approach to the voluntary sector', p. 127.

3. Lewis, 'The boundary between voluntary and statutory service'; Finlayson, 'A moving frontier', Deakin, 'The perils of partnership'.
4. N. Seddon, *Who Cares? How State Funding and Political Activism Change Charity* (London: Civitas, 2007), p. 46.
5. *Ibid.*, pp. 144–46, p. 159.
6. F. Prochaska, 'Voluntary action – renaissance or decline?', History and Policy website.
7. N. Deakin, 'Civil society and civil renewal' in NCVO, *Voluntary Action*, p. 30.
8. B. Knight and S. Robson, *The Value and Independence of the Voluntary Sector* (Newcastle: Centre for Research and Innovation in Social Policy and Practice [CENTRIS], 2007), p. 55.
9. J. Harris, 'Introduction – civil society in British history: paradigm or peculiarity?' in J. Harris (ed.), *Civil Society in British History: Ideas, Identities, Institutions* (Oxford: Oxford University Press, 2003), pp. 1–2.
10. R. Putnam, *Bowling Alone: The Collapse and Revival of American Community* (New York: Simon and Schuster, 2000), pp. 18–19.
11. For a detailed rejection of the idea that Britain experienced a similar decline in social capital to the US, see P.A. Hall, 'Social capital in Britain', *British Journal of Political Science*, 29, (1999), pp. 417–61.
12. Office of National Statistics, *Social Capital: A Review of the Literature* (London: ONS, 2001), p. 5, pp. 19–21.
13. N. Deakin, 'Civil society and civil renewal' in NCVO, *Voluntary Action: Meeting the Challenges of the 21st Century* (NCVO: London, 2005), pp. 25–28.
14. Strategy Unit/Cabinet Office, *Private Action Public Benefit: A Review of Charities and the Wider Not-For-Profits Sector* (London: Cabinet Office, 2002), p. 32.
15. *The Future Role of the Third Sector*, p. 17.
16. Stimson, 'Blair declares war'.
17. D. Best and A. Campbell, *Summary of the NTA's National Prescribing Audit* (London: National Treatment Agency, 2006), pp. 3–4.
18. See *Updated Drug Strategy*; *Tackling Drugs to Build a Better Britain* and *Tackling Drugs Together*, p. 23.
19. T. Carnwarth and C. Ford, 'Methadone challenged on its home turf: is there a worrying methadone backlash about?' *Drink and Drug News*, 8 May 2006, p. 9.
20. R. Yates, 'Unpleasant and petulant,' letter to *Drink and Drug News*, 22 May 2006, p. 8.
21. N. McKeganey, Z. Morris, J. Neale and M. Robertson, 'What are drug users looking for when they contact drug services: abstinence or harm reduction?' *Drugs: Education, Prevention and Policy*, 11, (2004), p. 426.
22. D. Best, A. Campbell and A. O'Grady, *The NTA's First Annual User Survey 2005* (London: NTA, 2006), p. 9.
23. For a detailed deconstruction of these figures, see M. Ashton, 'The new abstentionists', *Druglink special insert*, (Dec/Jan 2008), pp. 1–16.
24. BBC News, 'Drug services make slow progress', 30 October 2007, http://news.bbc.co.uk/go/pr/fr/-/1/hi/uk/7068572.stm, accessed 28 January 2008.
25. See for example, R. Cooke, 'Interview with David Cameron', *The Observer*, 11 February 2007, http://observer.guardian.co.uk/woman/story/0,,2007676,00.

html, accessed 30 January 2008; D. Cameron, Speech to the Annual Convention of the Youth Justice Board in Cardiff, Thursday November 2, 2006, http://www.conservatives.com/tile.do?def=news.story.page&obj_id= 133328&speeches=1, accessed 30 January 2008.

26. K. Gyngell, *Breakthrough Britain: Ending the Costs of Social Breakdown, Vol 4 – Addictions* (London: Social Justice Policy Group, 2007), p. 10, p. 13.

27. Interview by Virginia Berridge with NTA worker, 30 March, 2007.

28. See, for example, Cm 4818-I, *The NHS Plan: A Plan for Investment, A Plan for Reform* (London: TSO, 2000).

29. Source: http://www.drugs.gov.uk/drug-interventions-programme/, accessed 28 January 2008.

30. Home Office, *Tackling Drugs, Changing Lives: Keeping Communities Safe From Drugs* (London: Home Office, 2004), p. 19.

31. *Drugs Act, 2005.*

32. *Tackling Drugs, Changing Lives*, p. 19.

33. Finch and Ashton, 'Treatment to order', p. 189.

34. Kothari, Marsden and Strang, 'Effective treatment of drug misusers', p. 415.

35. G.M. Craig, 'Involving users in developing services', *British Medical Journal*, 336, (9 February 2008), pp. 286–87.

36. *The Future Role of the Third Sector*, p. 3.

37. See S. Rolles, 'The year is 2022 and drugs are legal', *Druglink*, (Jan/Feb 2008), pp. 6–9 and M. Hough, 'The new Puritanism', *Druglink*, (Jan/Feb 2008), pp. 14–15.

# Bibliography

## Primary Sources

### Archival Sources

*The National Archives hereafter (TNA)*

TNA FD 23/1949, Institute for the Study of Drug Dependence: proposal to set up the Institute, 1967–1976.

TNA HO 383/396, Drug addicts: Featherstone Lodge Project, SE London (Maudsley Hospital Institute of Psychiatry) minutes of meetings, news-sheets and reports. 1969–1972.

TNA HO 383/397, Drug addicts: Featherstone Lodge Project, SE London (Maudsley Hospital Institute of Psychiatry) minutes of meetings, news-sheets and reports, 1970–1974.

TNA HO 383/398, Drug addicts: Featherstone Lodge Project, SE London (Maudsley Hospital Institute of Psychiatry) minutes of meetings, news-sheets and reports, 1972–1980.

TNA MH 154/367, Drug dependence: rehabilitation and aftercare of addicts; sub-committee reports and recommendations, 1967–1971.

TNA MH 154/433, Community Drug Project: reports and requests for grants, 1970–1983.

TNA MH 154/430, Heroin addiction: London Boroughs Association; working party reports on rehabilitation, 1968–1974.

TNA MH 154/1148, Advisory Council on the Misuse of Drugs: Working Group on Treatment and Rehabilitation; papers, minutes of meetings and first interim report, 1977.

TNA MH 154/1149, Advisory Council on the Misuse of Drugs: Working Group on Treatment and Rehabilitation; preparation, handling and publication of the first interim report; draft Submission to Ministers, 1977.

TNA MH 154/1151, Advisory Council on the Misuse of Drugs: Working Group on Treatment and Rehabilitation; future programme of work arising from responses to the first interim report, 1978–1979.

TNA MH 154/1192, Future developments at SCODA, report to the DHSS, 1976–1977.

TNA MH 154/1251, Standing Conference on Drug Abuse (SCODA): proposals for a short stay residential centre for multiple drug abusers; opening of the City Roads (Crisis Intervention) Ltd, 358 City Road, London; funding arrangements, 1977.

TNA MH 154/1252, Standing Conference on Drug Abuse (SCODA): proposals for a short stay residential centre for multiple drug abusers; opening of the City Roads (Crisis Intervention) Ltd, 358 City Road, London; draft operational policy document, 1977.

TNA MH 154/1253, Standing Conference on Drug Abuse (SCODA): proposals for a short stay residential centre for multiple drug abusers; opening of the

City Roads (Crisis Intervention) Ltd, 358 City Road, London; draft operational policy document, 1977.

TNA MH 154/1254, Standing Conference on Drug Abuse (SCODA): proposals for a short stay residential centre for multiple drug abusers; opening of the City Roads (Crisis Intervention) Ltd, 358 City Road, London; operational policy document, 1971–1978.

TNA MH 154/1255, Standing Conference on Drug Abuse (SCODA): proposals for a short stay residential centre for multiple drug abusers; opening of the City Roads (Crisis Intervention) Ltd, 358 City Road, London; operational policy document, 1978.

TNA MH 154/1256, Standing Conference on Drug Abuse (SCODA): proposals for a short stay residential centre for multiple drug abusers; opening of the City Roads (Crisis Intervention) Ltd, 358 City Road, London; annual progress report, 1978.

TNA MH 154/1257, Standing Conference on Drug Abuse (SCODA): proposals for a short stay residential centre for multiple drug abusers; opening of the City Roads (Crisis Intervention) Ltd, 358 City Road, London; formal opening 18 October 1978; evaluation report, 1978.

TNA MH 154/1258, Standing Conference on Drug Abuse (SCODA): proposals for a short stay residential centre for multiple drug abusers; opening of the London Drug Crisis Centre, 358 City Road; review/finance meeting, 1979.

## Modern Records Centre, University of Warwick (hereafter MRC) Papers of Release

MRC MSS.171/1/2, Business Meeting and Policy Meeting Minutes, 1971–1972.

MRC MSS.171/1/5, Minutes, 1973–1974.

MRC MSS.171/2/1, End of year accounts, 1972–1975.

MRC MSS.171/3/12/11, Speeches, drafts, etc., 1967–1970.

MRC MSS.171/3/12/12, Publications, 1967–1971.

MRC MSS.171/3/12/13, Government Committees, 1968–1969, 1972.

MRC MSS.171/3/12/17, Trusts, appeals and mailing lists, 1970–1971.

MRC MSS.171/3/12/23, Release: A Fifth Anniversary Policy Statement, 1972.

MRC MSS.171/3/12/28, Miscellaneous, including Release discussion documents, 1975–1976.

MRC MSS.171/3/12/30, The Princedale Trust. Bi-Annual Report on the Work of Release, 1977.

MRC MSS.171/3/19/1, Donators and fund raising correspondence, 1969–1970.

MRC MSS.171/3/19/2, Donators and fund raising correspondence, 1971.

MRC MSS.171/3/44/1, Controversy regarding the GPO's decision to disconnect the service, 1975.

MRC MSS.171/4/7, Connection, 1972.

MRC MSS.171/4/8, Connection, 1972.

MRC MSS.171/4/9, Release Monthly Newsletter May, 1974.

MRC MSS.171/4/14, Release 67–77: Release 10th Anniversary Publication, 1977.

MRC MSS.171/4/15, News Release, 1977–1980.

MRC MSS.171/4/16, News Release, 1977–1980.

MRC MSS.171/4/17, News Release, 1977–1980.

MRC MSS.171/4/18, News Release, 1977–1980.

MRC MSS.171/4/19, News Release, 1977–1980.
MRC MSS.171/4/20, News Release, 1977–1980.
MRC MSS.171/4/21, News Release, 1977–1980.
MRC MSS.171/4/22, News Release, 1977–1980.
MRC MSS.171/4/23, News Release, 1977–1980.
MRC MSS.171/5/1/2, Release: A Progress Report, 1972.

*Department of Health Archive, Nelson, Lancashire (hereafter DOHA)*

DOHA OCG/1/1/3/VA/SSI/535, Treatment for drug misusers: Central Funding Initiative: Circular HN(83)13/LASSL (83)1, 1984–1985.
DOHA DAC/0026/V001, ACMD Report on Treatment and Rehabilitation Evaluation drug misuse central funding initiative, 1983.
DOHA DAC/0014/V0004, Advisory Council on Misuse of Drugs report on Treatment and Rehabilitation, Publication and Follow Up, 1983.
DOHA JR/0198/0565/V0001A, Research Personal Social Services and health: Assessment of the Centrally Funded Initiative on Services for Drug Misuses, Dr S MacGregor, 1985–1987.
DOHA JR/0198/0401, Advisory Council on the Misuse of Drugs: Report on Treatment and Rehabilitation, 1982–1983.
DOHA DAC/0007/V0004, Advisory Council on the Misuse of Drugs, TRWG [Treatment and Rehabilitation Working Group], 1982.
DOHA DAC0090/V002, SCODA/Drugwatch, 1989–1994.

*Contemporary Medical Archives Centre, Wellcome Library, London (hereafter CMAC), Private Papers of Dr Ann Dally (PP/DAL)*

CMAC PP/DAL/B/4/1/3/1, DDIG Minutes, 1987.
CMAC PP/DAL/B/4/1/3/2, DDIG correspondence, 1987.

*Private Papers of Professor Griffith Edwards (PP/GE)*

PP/GE File 3B.
PP/GE File 4.
PP/GE File 4A.
PP/GE File 5A.
PP/GE File 6A.
PP/GE File 7A.
PP/GE File 9B.
PP/GE File 11.
PP/GE File 18.
PP/GE File 19.

*DrugScope Library (DL)*

DL 14039, *The SCODA First Report* (London: SCODA, 1974).
DL 42638, *SCODA Annual Report, 1983–84*.
DL 48602, *SCODA Annual Report, 1985–86*.
DL 26430, Report of the conference 'Ten years after', held 1 July 1976.
DL 30658, *SCODA Annual Report, 1977–78*.
DL 32718, *SCODA Annual Report, 1978–79*.

DL 36959, *SCODA Annual Report, 1980–81.*
DL 29050, *SCODA Annual Report, 1976–77.*
DL 23221, *SCODA Annual Report, 1974–75.*

## Oral History Interviews

Interview with ADFAM worker by Virginia Berridge, 22 March 2007.
Interview with Tony Coxon, by Virginia Berridge, 15 March 1990.
Interview with a drug policy official, by Virginia Berridge, 24 January 2007.
Interview with a drug user activist, by Alex Mold, 27 July 2006.
Interview with a drug user activist by Alex Mold, 20 July 2006.
Interview with a drug user activist by Alex Mold, 13 April 2006.
Interview with a drug user activist, by Alex Mold, 13 December 2005.
Interview with a drug user activist, by Alex Mold, 19 July 2005.
Interview with a drug user activist, by Alex Mold, 23 August 2005.
Interview with a drug user activist, by Alex Mold, 6 July 2005.
Interview with a drug user activist, by Alex Mold, 30 May 2006.
Interview with Griffith Edwards, by Alex Mold, 20 October 2004.
Interview with Betsy Ettorre, by Virginia Berridge, February 1990.
Interview with General Practitioner with an interest in addiction treatment, by
   Virginia Berridge, 10 December 2006.
Interview with Janet Green by Janet Foster, AIDS Social History Programme,
   28 January 1993.
Interview with Jonathan Grimshaw, by Virginia Berridge, 2 March 1990.
Interview with Roger Howard, by Virginia Berridge, 20 March 2006.
Interview with a key civil servant by Alex Mold and Virginia Berridge, 2 May
   2006.
Interview with Kenneth Leech, by Alex Mold, 9 November 2004.
Interview with Jane MacGregor, by Alex Mold 22 March 2007.
Interview with National Treatment Association worker, by Virginia Berridge,
   30 March 2007.
Interview with Bill Nelles, by Virginia Berridge, 1 February 1990.
Interview with Bill Nelles, by Alex Mold and Virginia Berridge, 10 March 2006.
Interview with NTA official, by Virginia Berridge, 13 March 2007.
Interview with Andrew Preston by Virginia Berridge, 27 March 2007.
Interview with a psychiatrist connected to Phoenix House, by Alex Mold,
   11 May 2006.
Interview with a Sheffield based psychiatrist, by Alex Mold and Virginia Berridge,
   2 May 2006.
Interview with a social science researcher, by Virginia Berridge, 10 February
   2005.
Interview with Gerry Stimson, by Alex Mold and Virginia Berridge, 17 May 2006.
Interview with Sally Taylorson by Virginia Berridge, 13 April 2007.
Interview with David Tomlinson, by Alex Mold, 11 July 2006.
Interview conducted with senior Turning Point manager, by Virginia Berridge,
   12 March 2007.
Interview with Jasper Woodcock, by Alex Mold and Virginia Berridge,
   10 November 2004.
Interview with Rowdy Yates, by Alex Mold and Virginia Berridge, 10 March
   2006.

Interview with SCODA worker by Alex Mold, 25 February 2005.
Interview with Tony Whitehead, by Virginia Berridge and Phil Strong, 6 July 1989.

## Published Primary Sources

*Journals*

Druglink
The Hansard Journal of Parliamentary Debates: House of Commons
The Hansard Journal of Parliamentary Debates: House of Lords
The Lancet
British Medical Journal

*Reports*

Aves, G.M., *The Voluntary Worker in the Social Services* (London: Bedford Square Press, 1969).
Advisory Committee on Drug Dependence, *The Rehabilitation of Drug Addicts* (London: HMSO, 1968).
Advisory Committee on Drug Dependence, *Cannabis* (HMSO: London, 1968).
Advisory Committee on Drug Dependence, *Police Powers of Search and Arrest* (London: HMSO, 1969).
Advisory Council on the Misuse of Drugs, *Hidden Harm: Responding to the Needs of the Children of Problem Drug Users* (London: HMSO, 2003).
Advisory Council on the Misuse of Drugs, *Treatment and Rehabilitation* (London: HMSO, 1982).
Advisory Council on the Misuse of Drugs, *AIDS and Drug Misuse, Part One* (London: HMSO, 1988).
Blenheim Project, *People Adrift: The Work of the Blenheim Project With Young Drifters* (London: Blenheim Project, 1973).
Berridge, V., Christie, D.A. and Tansey, E.M. (eds), *Public Health in the 1980s and 1990s: Decline and Rise?* (London: Wellcome Centre, 2006).
Best, D. and Campbell, A., *Summary of the NTA's National Prescribing Audit* (NTA: London, 2006).
Best, D., A. Campbell and A. O'Grady, *The NTA's First Annual User Survey 2005* (London: NTA, 2006).
Beveridge, W., *Voluntary Action: A Report on Methods of Social Advance* (London: George Allen & Unwin, 1948).
Birchall, J. and Simmons, R., *User Power: The Participation of Users in Public Services* (London: National Consumer Council, 2004).
Branfield, F. and Beresford, P., *Making User Involvement Work: Supporting Service User Networking and Knowledge* (York: Joseph Rowntree Foundation, 2006).
British Medical Association, *The Misuse of Drugs* (London: Harwood Academic Publishers, 1997).
Coke Hole Trust, *The Coke Hole Trust 3rd Annual Report, 1975–76*.
Cm 1599, *The Citizen's Charter: Raising the Standard* (London: HMSO, 1991).
Cmd 2846, *Tackling Drugs Together: A Strategy for England 1995–1998* (London: The Stationary Office, 1995).
Cmd 3945, *Tackling Drugs to Build a Better Britain* (London: The Stationary Office, 1998).

Cm 4100, *Compact on Relations Between Government and the Voluntary and Community Sector in England* (London: HMSO, 1998).

Cmd 4221, *The Government's Expenditure Plans 1990–00 to 2001–02* (London: The Stationary Office, 1999).

Cm 4818-I, *The NHS Plan: A Plan for Investment, A Plan for Reform* (London: TSO, 2000).

Cm 6737, *Our Health, Our Care, Our Say: A New Direction for Community Health Services* (London: The Stationary Office, 2006).

Department of Health, *Working for Patients* (London: HMSO, 1989).

Department of Health, *NHS Plan: A Plan for Investment, A Plan for Reform* (London: The Stationery Office, 2000).

Department of Health, *The Expert Patient: A New Approach to Chronic Disease Management for the 21st Century* (London: The Stationary Office, 2001).

Department of Health, *Statistics from the National Drug Treatment Monitoring System (NDTMS) 1 April 2003–31 March 2004* (London: Department of Health, 2005).

Department of Health, *Choosing Health: Making Healthy Choices Easier* (London: The Stationary Office, 2004).

Department of Health *Our Health, Our Care, Our Say* (London, Department of Health, 2006).

Department of Health, *Choice Matters 2007–08: Putting Patients in Control,* (London: Department of Health, 2007).

Department of Health, *NHS Choices: Delivering for the NHS,* (London, Department of Health, 2008).

Department of Health, *NHS Constitution* (London: Department of Health, 2009).

Department of Health, *Drug Misuse and Dependence: Guidelines on Clinical Management – Update June 2007, Consultation Draft* (London: Department of Health, 2007).

European Association for the Treatment of Addiction, *Service User Views of Treatment: Research Conducted for the Audit Commission* (London: EATA, 2004).

Fischer, J., Jenkins, N., Bloor, M., Neale, J. and Berney, L., *Drug User Involvement in Treatment Decisions* (York: Joseph Rowntree Foundation, 2007).

Gossop, M., *Drug Misuse Treatment and Reductions in Crime: Findings From The National Treatment Outcome Research Study (NTORS)* (London: NTA, 2005).

Griffiths, R., *Community Care: An Agenda For Action: A Report to the Secretary of State for Social Services* (London: HMSO, 1988).

Gyngell, K., *Breakthrough Britain: Ending the Costs of Social Breakdown, Vol 4 – Addictions* (London: Social Justice Policy Group, 2007).

HM Treasury, *Exploring the Role of the Third Sector in Public Service Delivery and Reform: A Discussion Document* (London: HM Treasury, 2004).

HM Treasury, Cm 7189: *The Future Role of the Third Sector in Social and Economic Regeneration: Final Report* (London: The Stationary Office, 2007).

Home Office, *Tackling Drug Misuse: A Summary of the Government's Strategy* (London: HMSO, 1985).

Home Office, *Statistics of Drug Addicts Notified to the Home Office, 1988* (London: HMSO, 1989).

Home Office, *Efficiency Scrutiny of Government Funding of the Voluntary Sector: Profiting from Partnership* (London: Home Office, 1990).

Home Office, *Updated Drug Strategy* (London: HMSO, 2002).

Home Office, *2003 Home Office Citizenship Survey: People, Families and Communities* (London, Home Office: 2004).

Home Office, *Tackling Drugs, Changing Lives: Keeping Communities Safe From Drugs* (London: Home Office, 2004).

Home Office and Department of Health, *Review of Second Tier Activity in the Drugs Sector: Final Report* (London: Cordis Bright, 2006).

House of Commons Social Services Committee, *Misuse of Drugs With Special Reference to the Treatment and Rehabilitation of Misusers of Hard Drugs, Session 1984–1985* (London: HMSO, 1984–1986).

Jamieson, A., Glanz, A. and MacGregor, S., *Dealing With Drug Misuse: Crisis Intervention in the City* (London: Tavistock, 1984).

Joseph Rowntree Trust, *The Future of Voluntary Organisations: Report of the Wolfenden Committee* (London: Croom Helm, 1978).

Knight, B., *Voluntary Action* (London: Home Office, 1993).

Knight B. and Robson, S. *The Value and Independence of the Voluntary Sector* (Newcastle: Centre for Research and Innovation in Social Policy and Practice (CENTRIS), 2007).

MacGregor, S. et al., *Drug Services in England and the Impact of the Central Funding Initiative* (London: ISDD, 1991).

McDermott, P., *Out-patient Treatment for Heroin Addiction: A Service Users' Guide to Rights and Responsibilities* (Manchester: Lifeline Publications, 2003).

Ministry of Health, *Report of the Working Party on Social Workers in Local Authority Health and Welfare Services* (London: HMSO, 1959).

Ministry of Health, *Drug Addiction: Second Report of the Interdepartmental Committee* (London: HMSO, 1965).

National Council for Voluntary Organisations, *The Voluntary Sector Almanac 2004* (London: NCVO, 2004).

National Council for Voluntary Organisations, *Voluntary Action: Meeting the Challenges of the 21st Century* (NCVO: London, 2005).

National Council for Voluntary Organisations, *The UK Voluntary Sector Almanac 2006* (London: NCVO, 2006).

National Treatment Agency, *Supporting and Involving Carers* (London: NTA, 2006).

National Treatment Agency, *NTA Guidance For Local Partnerships on User and Carer Involvement* (London: NTA, 2006).

National Treatment Agency, *The NTA's First Annual User Satisfaction Survey 2005* (London: NTA, 2006).

National Treatment Agency, *Being Heard: Notable Examples of User and Carer Organisations* (NTA: London, 2004).

Office of National Statistics, *Social Capital: A Review of the Literature* (London: ONS, 2001).

Royal College of General Practitioners, *Guidance for the use of Buprenorphine for the Treatment of Opioid Dependence in Primary Care* (London: RCGP, 2004).

Royal College of General Practitioners, *Guidance for the use of Methadone for the Treatment of Opioid Dependence in Primary Care* (London: RCGP, 2005).

SCODA, *Drug Problems: Where to Get Help* (London: SCODA, 1986).

SCODA, *Getting Drug Users Involved, Good Practice in Local Treatment and Planning* (London: SCODA, 1997).

Scottish Home and Health Department, *HIV Infection in Scotland: the Report of the Scottish Committee on HIV Infection and Intravenous Drug Misuse* (Edinburgh: Scottish Home and Health Department, 1986).

Seddon, N., *Who Cares? How State Funding and Political Activism Change Charity* (London: Civitas, 2007).

Social Enterprise Coalition, *There's More to Business Than You Think: A Guide to Social Enterprise* (2003).

Southwell, M., *A Guide to Involving and Empowering Drug Users*, no date [2003?], http://www.harmreductionnetwork.mb.ca/nta.pdf, accessed 23 May 2007.

Strategy Unit/Cabinet Office, *Private Action Public Benefit: A Review of Charities and the Wider Not-For-Profits Sector* (London: Cabinet Office, 2002).

Wanless, D., *Securing our Future Health: Taking a Long Term View* (London: HM Treasury, 2002).

### Primary Books and Articles

Ashton, M., 'Radical change risks crisis of confidence in SCODA', *Druglink*, 10:1, (Jan/Feb 1995), pp. 5–8.

Ashton, M., 'The new abstentionists', *Druglink special insert*, (Dec/Jan 2008), pp. 1–16.

Barnes, M. 'Back to the roots,' *Druglink*, (Jan/Feb 2007), pp. 8–9.

Burr, A., "Increased sale of opiates on the blackmarket [sic.] in the Piccadilly area," *British Medical Journal*, (24 September 1983), pp. 883–85.

Cave, M. 'Fighting for survival', *Drink and Drug News*, (8 May 2006), p. 12.

Carnwarth, T. and C. Ford, 'Methadone challenged on its home turf: is there a worrying methadone backlash about?' *Drink and Drug News*, (8 May 2006), p. 9.

Clement, S. and Strang, J., 'The rise and fall of the Community Drugs team: the gap between aspiration and achievement' in J. Strang and M. Gossop (eds), *Heroin Addiction and the British System Volume 2* (London: Routledge, 2005), pp. 94–104.

Coon, C. and Harris, R., *The Release Report On Drug Offenders and the Law* (London: Sphere Books, 1969).

Coon, C., 'We were the welfare branch of the alternative society' in Curtis and Sanderson (eds), *Unsung Sixties*, pp. 183–97.

Craig, G.M., 'Involving users in developing services', *British Medical Journal*, 336, (9 February 2008), pp. 286–87.

Dale Perera, A., 'National Treatment Agency. Building castles on sand?' *Druglink*, (May/June 2001), pp. 19–21.

Duke, K., 'Out of crime and into treatment? The criminalization of contemporary drug policy since *Tackling Drugs Together*', *Drugs: Education, Prevention and Policy*, 13:5, (2006), pp. 409–415.

Edwards, G., 'Relevance of American experience of narcotic addiction to the British scene', *British Medical Journal*, (12 August 1967), pp. 425–429.

Farley, P., 'Can these bones live?' *Druglink*, 14:1, (Jan/Feb1999), pp. 18–19.

Finch, E. and Ashton, M., 'Treatment to order: the new drug treatment and testing orders' in J. Strang and M. Gossop (eds), *Heroin Addiction and the British System Volume 2* (London: Routledge, 2005), pp. 187–197.

Gay, P., *Getting Together: A Study of Self-Help Groups for Drug Users Families* (London: PSI Publications, 1989).

Ghodse, A.H., 'Casualty departments and the monitoring of drug dependence', *British Medical Journal*, 1, (28 May 1977), pp. 1381–82.

Ghodse, A.H., 'Conversation with Hamid Ghodse', *Addiction*, 102, (2007), pp. 197–205.

Green, J., *Days in the Life: Voices From the English Underground* (London: Pimlico, 2nd edition, 1988).

Harrison, S. and Mort, M. 'Which champions, which people? Public and user involvement in health care as a technology of legitimation', *Social Policy and Administration*, 32:1, (1998), pp. 60–70.

Hough, M., 'Balancing public health and criminal justice interventions', *International Journal of Drug Policy*, 12, (2001), pp. 429–433.

Hough, M., 'The new Puritanism', *Druglink*, (Jan/Feb 2008), pp. 14–15.

Kothari, G., Marsden, J. and Strang, J., 'Opportunities and obstacles for effective treatment of drug misusers in the criminal justice system in England and Wales', *British Journal of Criminology*, 42, (2002), pp. 412–432.

Leech, K., *Pastoral Care and the Drug Scene* (London: SPCK, 1970).

Leech, K., 'The role of voluntary agencies in the various aspects of prevention and rehabilitation of drug abusers', *British Journal of Addiction*, 67, (1972), pp. 131–36.

Leech, K., *Care and Conflict: Leaves From A Pastoral Notebook* (London: Darton, Longman & Todd: 1990).

Logan, F. (ed.), *Cannabis: Options For Control* (London: ISDD, 1979).

Love, J. and Gossop, M., 'The processes of referral and disposal within a London Drug Dependence Clinic', *British Journal of Addiction*, 80, (1985), pp. 435–40.

McDermott, P., 'User run, user friendly: why drug workers should empower drug users into their jobs', *Druglink*, (November/December, 1992), p. 11.

McKeganey, N., Z. Morris, J. Neale and M. Robertson, 'What are drug users looking for when they contact drug services: abstinence or harm reduction?' *Drugs: Education, Prevention and Policy*, 11, (2004), pp. 423–35.

Nutley, S.M., I. Walter and H.T.O. Davies, *Using Evidence. How Research Can Inform Public Services* (Bristol: The Policy Press, 2007).

Nutley, S. M., N. Bland and I. C. Walter, 'The institutional arrangements for connecting evidence and policy: the case of drug misuse', *Public Policy and Administration*,17:3, (2002), pp.76–94.

O'Mara, E., 'Opinion: Five years ago there was there was a promise of better things to come through "user involvement". In hindsight have those promises been fulfilled? No.', *Drugs and Alcohol Today*, 4:1, (2004), pp. 18–19.

Ogbourne, A., 'The first 100 residents in a therapeutic community for former addicts', *British Journal of Addiction*, 60, (1975), pp. 65–76.

Ogbourne, A. and Melotte, C., 'An evaluation of a therapeutic community for former drug users', *British Journal of Addiction*, 72, (1977), pp. 75–82.

Parker, H., Newcombe, R. and Bakx, K., 'The new heroin users: prevalence and characteristics in Wirral, Merseyside', *British Journal of Addiction*, 82, (1987), pp. 147–57.

Parker, H., K. Bakx and R. Newcombe, *Living with Heroin. The Impact of a Heroin Epidemic in an English Community* (Buckingham: Open University Press, 1988).

Patterson, S., T. Weaver, K. Agath, E. Albert and T. Rhodes, ' "They can't solve the problem without us": a qualitative study of stakeholder perspectives on user involvement in drug treatment services in England', *Health and Social Care in the Community*, 17:1, (2008), pp. 54–62.

Peck, D.F. and Plant, M.A., 'Unemployment and illegal drug use: concordant evidence from a prospective study and national trends', *British Medical Journal*, 293, (11 October 1986), pp. 929–32.

Polkinghorne, J., M. Gossop and J. Strang, 'The Government Task Force and its review of drug treatment services. The promotion of an evidence based approach' in J. Strang and M. Gossop (eds), *Heroin Addiction and the British System Volume 2* (London: Routledge, 2005), pp.198–205.

Robertson, J.R., A.B.V. Bucknall, P.D. Welsby et al., 'Epidemic of AIDS related virus (HTLVIII/LAV) infection among intravenous drug users', *British Medical Journal*, 292, (1986), p. 527.

Rolles, S., 'The year is 2022 and drugs are legal', *Druglink*, (Jan/Feb 2008), pp. 6–9.

Shapiro, H., 'Nothing about us, without us: user involvement, past, present and future', *Druglink*, (May/June 2005), pp. 10–11.

Smiles, S., *Self-Help: With Illustrations of Character, Conduct and Perseverance* (London: IEA, 1997, first published 1859).

Spear, H.B. 'The growth of heroin addiction in the United Kingdom', *British Journal of Addiction*, 64, (1969), pp. 245–55.

Stimson, G. and Oppenheimer, E., *Heroin Addiction: Treatment and Control in Britain* (London: Tavistock, 1982).

Stimson, G., 'Blair declares war the unhealthy state of British drug policy', *International Journal of Drug Policy*, 11, (2000), pp. 259–64.

Strang, J. 'A model service: turning the generalist on to drugs' in S. MacGregor (ed.), *Drugs and British Society* (London: Routledge, 1989).

Strang, J. 'The roles of prescribing' in J. Strang, and G.V. Stimson (eds), *AIDS and Drug Misuse: The Challenge for Policy and Practice in the 1990s* (London: Routledge, 1990), pp. 142–52.

Whitehead, T., 'The voluntary sector: Five years on' in E. Carter and S. Watney (eds), *Taking Liberties: AIDS and Cultural Politics* (London: Serpent's Tail, 1989), pp. 107–11.

Woodcock, J., 'The growth of information: the development of Britain's nation drug misuse information resource' in J. Strang and M. Gossop (eds), *Heroin Addiction and Drug Policy: The British System* (Oxford: Oxford University Press, 1994), pp. 304–310.

Young, J., *The Drugtakers: The Social Meaning of Drug Use* (London: Paladin, 1972).

## Secondary Sources

6, P. and D. Leat, 'Inventing the British voluntary sector by committee: from Wolfenden to Deakin', *Non Profit Studies*, 1, (1996), pp. 33–45.

6, P., 'Giving consumers of British public services more choice: what can be learned from recent history?' *Journal of Social Policy*, 32:2, (2003), pp. 239–70.

Andersson, E., Creasy, S. and Tritter, J., 'Overview: does patient and public involvement matter?' in E. Andersson, J. Tritter and R. Wilson (eds), *Healthy*

*Democracy: The Future of Involvement in Health and Social Care* (London: Involve and NHS Centre for Involvement, 2007), pp. 7–17.

Anker, J., 'Active drug users – struggling for rights and recognition' in J. Anker, V. Asmussen, P. Kouvonen and D. Tops (eds), *Drug Users and Spaces for Legitimate Action* (Helsinki: Nordic Council for Drug and Alcohol Research, 2006), pp. 37–60.

Anker, J., Asmussen, V., Kouvonen, P. and Tops, D., 'Introduction' in J. Anker, V. Asmussen, P. Kouvonen and D. Tops (eds), *Drug Users and Spaces for Legitimate Action* (Helsinki: Nordic Council for Drug and Alcohol Research, 2006), pp. 5–22.

Asmussen, V., 'User participation in Danish methadone maintenance treatment' in J. Anker, V. Asmussen, P. Kouvonen and D. Tops (eds), *Drug Users and Spaces for Legitimate Action* (Helsinki: Nordic Council for Drug and Alcohol Research, 2006), pp. 205–26.

Baggott, R., Allsop, J. and Jones, K., *Speaking for Patients and Carers: Health Consumer Groups and the Policy Process* (Basingstoke: Palgrave, 2005).

Bagguley, P., 'Social change, the middle classes and the emergence of "new social movements": a critical analysis', *The Sociological Review*, 40:1, (1992), pp. 26–48.

Barnes, M., Harrison, S., Mort, M. and Shardlow, P., *Unequal Partners: User Groups and Community Care* (Bristol: Policy Press, 1999).

Bennett, D., 'The drive towards the community' in G.E. Berrios and H. Freeman (eds), *150 Years of British Psychiatry, 1841–1991* (London: Royal College of Psychiatrists, 1991), pp. 321–32.

Berridge, V., 'AIDS and British drug policy: continuity or change?' in V. Berridge and P. Strong (eds), *AIDS and Contemporary History* (Cambridge: Cambridge University Press, 1993), pp. 135–56.

Berridge, V., *AIDS in the UK: The Making of Policy, 1981–1994* (Oxford: Oxford University Press, 1996).

Berridge, V., ' "Unambiguous voluntarism?" AIDS and the voluntary sector in the United Kingdom' in C. Hannaway (ed.), *AIDS and the Public Debate* (Amsterdam: IOS Press, 1995), pp. 153–69.

Berridge, V., *Opium and the People: Opiate Use and Drug Control Policy in Nineteenth and Early Twentieth Century England* (Revised edition, London: Free Association Books, 1999).

Berridge, V., *Health and Society in Britain Since 1939* (Cambridge: Cambridge University Press, 1999).

Berridge, V., 'Punishment or treatment? Inebriety, drink and drugs, 1860–1914', *Lancet*, 364, (2004), pp. 4–5.

Berridge, V., *Marketing Health. Smoking and the Discourse of Public Health in Britain, 1945–c.2000* (Oxford: Oxford University Press, 2007).

Berridge, V., 'The "British system" and its history: myth and reality' in J. Strang and M. Gossop (eds), *Heroin Addiction and the British System. Volume 1 Origins and Evolution* (London: Routledge, 2005), pp. 7–16.

Berridge, V., 'New social movement or government funded voluntary sector? ASH, (Action on Smoking and Health) science and anti-tobacco activism in the 1970s' in M. Pelling and S. Mandelbrote (eds), *The Practice of Reform in Health, Medicine and Science, 1500–2000. Essays for Charles Webster* (London: Ashgate, 2005), pp. 333–48.

Beresford, P. and Wilson, A., 'Genes spell danger: mental health service users/survivors, bioethics and control', *Disability and Society*, 17:5, (2002), pp. 541–53.

Black, N., 'The NHS Research and Development Programme: The first five years, 1991–6', seminar paper at London School of Hygiene and Tropical Medicine, 21 January 1997.

Billis, D., *Organising Public and Voluntary Agencies* (London: Routledge, 1993).

Brenton, M., *The Voluntary Sector in British Social Services* (London: Longman, 1985).

Buchanan, T., ' "The truth will set you free": the making of Amnesty International', *Journal of Contemporary History*, 37:4, (2002), pp. 575–97.

Buchanan, T., 'Amnesty International in crisis: 1966–7', *Twentieth Century British History*, 15:3, (2004), pp. 267–89.

Butler, J., *Patients, Policies and Politics: Before and After Working for Patients* (Buckingham: Open University Press, 1992).

Byrne, P., *The Campaign For Nuclear Disarmament* (London: Croom Helm, 1988).

Byrne, P., *Social Movements in Britain* (London: Routledge, 1997).

Calhoun, C., ' "New social movements" of the early nineteenth century', *Social Science History*, 17:3, (1993), pp. 385–427.

Campling, P., 'Therapeutic communities', *Advances in Psychiatric Treatment*, 7, (2001), pp. 365–72.

Cotgrove, S. and Duff, A., 'Environmentalism, middle class radicalism and politics', *Sociological Review*, 28:2, (1980), pp. 333–51.

Crofts, N., 'A history of peer-based drug user groups in Australia', *Journal of Drug Issues*, 25, (1993), pp. 599–616.

Crossley, N., 'R.D. Laing and the British anti-psychiatry movement: a sociohistorical analysis', *Social Science and Medicine*, 47, (1998), pp. 877–89.

Crossley, N., 'Transforming the mental health field: the early history of the National Association for Mental Health', *Sociology of Health and Illness*, 20:4, (1998), pp. 458–88.

Crossley, N., 'Fish, field, habitus and madness: on the first wave of mental health users movement in Great Britain', *British Journal of Sociology*, 50:4, (1999), pp. 647–60.

Crossley, N. and Crossley, M., 'Patient voices, social movements and habitus. How psychiatric survivors speak out', *Social Science and Medicine*, 52, (2001), pp. 1477–89.

Crossely, N., *Making Sense of Social Movements* (Buckingham: Open University Press, 2002).

Crossley, N., *Contesting Psychiatry: Social Movements in Mental Health* (London: Routledge, 2005).

Curtis H. and Sanderson, M. (eds), *The Unsung Sixties: Memoirs of Social Innovation* (London: Whiting & Birch, 2004).

Daunton, M. (ed.), *Charity, Self-Interest and Welfare in the English Past* (London: UCL, 1996).

Davenport-Hines, R., *The Pursuit of Oblivion: A Global History of Narcotics, 1500–2000* (London: Weinfield & Nicholson, 2001).

Davis Smith, J., 'Philanthropy and self-help in Britain, 1500–1945' in J. Davis Smith, C. Rochester and R. Hedley (eds), *An Introduction to the Voluntary Sector* (London: Routledge, 1995), pp. 9–39.

Deakin, N., 'The perils of partnership: the voluntary sector and the state 1945–1992', in J. Davis Smith, C. Rochester and R. Hedley (eds), *An Introduction to the Voluntary Sector* (London: Routledge, 1995), pp. 40–65.

Deakin, N., *In Search of Civil Society* (Basingstoke: Palgrave, 2001).

Deakin, N., 'Civil society and civil renewal' in NCVO, *Voluntary Action: Meeting the Challenges of the 21st Century* (NCVO: London, 2005).

Dorn N. and South, N., *Helping Drug Users: Social Work, Advice Giving, Referral and Training Services of Three London Street Agencies* (Aldershot: Gower, 1985).

Duke, K., 'Out of crime and into treatment? The criminalization of contemporary drug policy since *Tackling Drugs Together*', *Drugs: Education, Prevention and Policy*, 13:5, (2006), pp. 409–15.

Eley, G., *Forging Democracy: The History of the Left in Europe, 1850–2000* (Oxford University Press: Oxford, 2002).

Epstein, S., *Impure Science. AIDS, Activism and the Politics of Knowledge* (Berkeley: University of California Press, 1996).

Evans, T. Stopping the poor getting poorer: the establishment and professionalisation of the poverty lobby, 1945–1995' in N. Crowson, M. Hilton and J. McKay (eds), *NGOs in Contemporary Britain: Non-State Actors in Society and Politics Since 1945* (Basingstoke: Palgrave Macmillan, 2009), pp. 147–63..

Finch, E. and Ashton, M., 'Treatment to order: the new drug treatment and testing orders' in J. Strang and M. Gossop (eds), *Heroin Addiction and the British System Vol 2* (London & New York: Routledge, 2005), pp. 187–97.

Finlayson, G., 'A moving frontier: voluntarism and the state in British social welfare, 1911–1949', *Twentieth Century British History*, 1:2, (1990), pp. 181–206.

Finlayson, G., *Citizen, State and Social Welfare in Britain, 1830–1990* (Oxford: Oxford University Press, 1994).

Forster, R. and J. Gabe, 'Voice or choice? Patient and public involvement in the National Health Service in England under New Labour', *International Journal of Health Services*, 38:2, (2008), pp. 333–56.

Foucault, M., *Madness and Civilisation: A History of Insanity in the Age of Reason* (London: Routledge, 1st English edition., 1967, this edition, 2001).

Foucault, M., *The Birth of the Clinic: An Archaeology of Medical Perception* (London: Tavistock Publications, 1973).

Friedman, S.R., des Jarlais, D.C. et al., 'AIDS and self organisation among intravenous drug users', *International Journal of the Addictions*, 22:3, (1987), pp. 210–19.

Friedman S.R., and Casriel, C., 'Drug users' organisations and AIDS Policy' *AIDS and Public Policy Journal*, 3:2, (1988), pp. 30–36.

Friedman, S.R., de Jong, W.M. and des Jarlais, D.C., 'Problems and dynamics of organising intravenous drugs users for AIDS prevention', *Health Education Research*, 3:1, (1988), pp. 49–57.

Friedman, S.R., 'Theoretical bases for understanding drug users' organisations', *International Journal of Drug Policy*, 7:4, (1996), pp. 212–19.

Friedman, S.R., 'The political economy of drug user scapegoating – the philosophy and politics of resistance', *Drugs: Education, Prevention and Policy*, 5:1, (1998), pp. 15–32.

Freeman, J. and Johnson, V. (eds), *Waves of Protest: Social Movements Since the Sixties* (Lanham, Maryland: Rowman and Littlefield, 1999).

Griffiths, P., M. Gossop and J. Strang, 'Chasing the dragon: the development of heroin smoking in the United Kingdom', in Strang and Gossop (eds), *Heroin Addiction and Drug Policy*, pp. 121–133.

Green, J., *All Dressed Up: The Sixties and the Counter-Culture* (London: Pimlico, 1999).

Greener, I., 'Who choosing what? The evolution of the use of "choice" in the NHS and its importance for New Labour' in C. Bochel, N. Ellison and M. Powell (eds), *Social Policy Review 15: UK and International Perspectives* (Bristol: The Social Policy Press, 2003), pp. 49–68.

Gorsky, M. and J. Mohan, *Don't Look Back? Voluntary and Charitable Finance of Hospitals in Britain, Past and Present* (London: Office of Health Economics and Chartered Accountants, 2001).

Gorsky, M. and S. Sheard (eds), *Financing Medicine: The British Experience Since 1750* (London: Routledge, 2006).

Gorsky, M., J. Mohan and T. Willis, *Mutualism and Health Care: British Hospital Contributory Schemes in the Twentieth Century* (Manchester: Manchester University Press, 2006).

Habermas, J., 'New social movements', *Telos*, 49, (1981), pp. 33–37.

Hall, P.A., 'Social capital in Britain', *British Journal of Political Science*, 29, (1999), pp. 417–61.

Ham, C.J., 'Power, patients and pluralism' in K. Barnard and K. Lee (eds), *Conflicts in the National Health Service* (London: Croom Helm, 1977), pp. 99–110.

Harris, J., 'Tradition and transformation: society and civil society in Britain, 1945–2000' in K. Burk (ed.), *The British Isles Since 1945* (Oxford: Oxford University Press, 2003), pp. 91–125.

Harris, J., 'Introduction – civil society in British history: paradigm or peculiarity?' in J. Harris (ed.), *Civil Society in British History: Ideas, Identities, Institutions* (Oxford: Oxford University Press, 2003), pp. 1–2.

Harris, M., C. Rochester and P. Halfpenny, 'Voluntary organisations and social policy: twenty years of change' in M. Harris and C. Rochester (eds), *Voluntary Organisations and Social Policy in Britain: Perspectives on Change and Choice* (Basingstoke: Palgrave, 2001), pp. 1–20.

Harrison, T., *Bion, Rickman, Foulkes and the Northfield Experiments* (London: Jessica Kingsley, 2000).

Hilton, M., *Consumerism in Twentieth-Century Britain: The Search for a Historical Movement* (Cambridge: Cambridge University Press, 2003).

Hogg, C., *Patients, Power and Politics: From Patients to Citizens* (London: Sage, 1999).

Illich, I., *Limits to Medicine – Medical Nemesis: The Expropriation of Health* (London: Marion Boyars, 1976).

Janzen, R., *The Rise and Fall of Synanon: A California Utopia* (Baltimore: Johns Hopkins University Press, 2001).

Jefferys, M., 'The transition from public health to community medicine', *Society for the Social History of Medicine Bulletin*, 39, (1986), pp. 47–63.

Jones, K., *Asylums and After: A Revised History of the Mental Health Services From the Early Eighteenth Century to the 1990s* (London: Athlone Press, 1993).

Johnson, N. 'The changing role of the voluntary sector in Britain from 1945 to the present day' in S. Kunhle and P. Selle (eds), *Government and Voluntary Organisations* (Aldershot: Ashgate, 1992), pp. 87–107.

Kendall, J. and Knapp, M., *The Voluntary Sector in the UK* (Manchester: Manchester University Press, 1996).

Kendall, J. and Knapp, M. 'A loose and baggy monster: boundaries, definitions and typologies' in J. Davis Smith, C. Rochester and R. Hedley (eds), *An Introduction to the Voluntary Sector* (London: Routledge, 1995), pp. 66–90.

Kendall, J., 'The mainstreaming of the third sector into public policy in England in the late 1990s: whys and wherefores', Civil Society Working Paper 2, January 2000, http://www.lse.ac.uk/collections/CCS/pdf/CSWP/cswp2.pdf, accessed 28 July 2009.

Kennard, D., *An Introduction to Therapeutic Communities* (London: Jessica Kingsley, 1998).

Kohn, M., *Narcomania: On Heroin* (London: Faber & Faber, 1987).

Kuhnle, S. and Selle, P., 'Government and voluntary organisations: a relational perspective' in S. Kuhnle and P. Selle (eds), *Government and Voluntary Organisations: A Relational Perspective* (Aldershot: Ashgate, 1992), pp. 1–33.

Kooyman, M., 'The history of therapeutic communities: a view from Europe' in B. Rawlings and R. Yates, *Therapeutic Communities for the Treatment of Drug Users* (London: Jessica Kingsley, 2001), pp. 59–78.

Laanemets, L., 'Organisation among drug users in Sweden' in J. Anker, V. Asmussen, P. Kouvonen and D. Tops (eds), *Drug Users and Spaces for Legitimate Action* (Helsinki: Nordic Council for Drug and Alcohol Research, 2006), pp. 105–29.

Lowe, R., *The Welfare State in Britain Since 1945* (3rd edition, Basingstoke: Palgrave, 2005).

Lewis, J., 'Developing the mixed economy of care: emerging issues for voluntary organisations', *Journal of Social Policy*, 22:2, (1993), pp. 173–92.

Lewis, J., 'The boundary between voluntary and statutory social service in the late nineteenth and early twentieth centuries', *Historical Journal*, 39, (1996), pp. 155–77.

Lewis, R., 'Flexible hierarchies and dynamic disorder – the trading and distribution of illicit heroin in Britain and Europe, 1970–1990' in Strang and Gossop (eds), *Heroin Addiction and Drug Policy*.

MacGregor, S., 'The public debate in the 1980s' in S. MacGregor (ed.), *Drugs and British Society: Responses to a Social Problem in the 1980s* (London: Routledge, 1989), pp. 1–19.

MacGregor, S. and Pimlott, B., 'Action and inaction in the cities' in S. MacGregor and B. Pimlott, *Tackling the Inner Cities: The 1980s Reviewed, Prospects for the 1990s* (Oxford: Clarendon Press, 1990).

Mäkelä, K., *Alcoholics Anonymous as a Mutual-Help Movement: A Study in Eight Societies* (Madison: University of Wisconsin Press, 1996).

Manning, N., *The Therapeutic Community Movement: Charisma and Routinization* (London: Routledge, 1989).

Mars, S., 'Peer pressure and imposed consensus: the making of the 1984 Guidelines of Good Clinical Practice in the Treatment of Drug Misuse' in V. Berridge (ed.), *Making Health Policy: Networks in Research and Policy After 1945* (Amsterdam: Rodopi, 2005).

Marwick, A., *The Sixties: Cultural Revolution in Britain, France, Italy and the United States, c.1968–c.1974* (Oxford: Oxford University Press, 1998).

Mattausch, J., *A Commitment to Campaign: A Sociological Study of CND* (Manchester: Manchester University Press, 1989).

Melucci, A. *Nomads of the Present* (London: Radius, 1988).

Mohan J. and Gorsky, M., *Don't Look Back? Voluntary and Charitable Finance of Hospitals in Britain, Past and Present* (London: Office of Health Economics, 2001).

Mold, A., 'The "British System" of heroin addiction treatment and the opening of the Drug Dependence Units, 1965–1970', *Social History of Medicine*, 17:3, (2004), pp. 501–17.

Mold, A., '"The welfare branch of the alternative society"? The work of drug voluntary organisation Release, 1967–1978', *Twentieth Century British History*, 17:1, (2006), pp. 50–73.

Mold, A., and Berridge, V., 'Crisis and opportunity in drug policy: changing the direction of British drug services in the 1980s', *Journal of Policy History*, 19:1, (2007), pp. 29–48.

Mold, A., *Heroin: The Treatment of Addiction in Twentieth-Century Britain* (De Kalb: Northern Illinois University Press, 2008).

Mold, A., 'Patient groups and the construction of the patient as "consumer" in Britain since the 1960s', unpublished seminar paper for the School of Population & Health Sciences Seminar, University of Birmingham 6 May 2009.

Mott, J., 'Notification and the Home Office' in J. Strang and M. Gossop (eds), *Heroin Addiction and Drug Policy: The British System* (Oxford: Oxford University Press, 1994), pp. 271–291.

Nicholson, M. and Lewis, G.W., 'The early history of the Multiple Sclerosis Society of Great Britain and Northern Ireland: a socio-historical study of lay/practitioner interaction in the context of a medical charity', *Medical History*, 46, (2002), pp. 141–74.

Offe, C., 'New social movements: challenging the boundaries of institutional politics', *Social Research*, 52:4, (1985), pp. 817–68.

Owen, D., *English Philanthropy 1660–1960* (Cambridge, Mass.; Harvard University Press, 1965).

Parkin, F., *Middle Class Radicalism: The Social Basis of CND* (Manchester: Manchester University Press, 1968).

Pearce, J.L. *Volunteers: The Organisational Behaviour of Unpaid Workers* (London: Routledge, 1993).

Pearson, G., and Gilman, M., 'Drug epidemics in space and time: local diversity, subcultures and social exclusion' in J. Strang and M. Gossop (eds), *Heroin Addiction and the British System: Volume 1* (London: Routledge, 2005), pp. 109–14.

Perkin, H., *The Rise of Professional Society: England Since 1880* (London: Routledge, 1989).

Power, R., 'Drug trends since 1968' in J. Strang and M. Gossop (eds), *Heroin Addiction and Drug Policy: The British System* (Oxford: Oxford University Press, 1996), pp. 27–41.

Prochaska, F., *Women and Philanthropy in Victorian England* (Oxford: Oxford University Press, 1980).

Prochaska, F. *The Voluntary Impulse: Philanthropy in Modern Britain* (London: Faber & Faber, 1988).

Prochaska, F., *Christianity and Social Service in Modern Britain: The Disinherited Spirit* (Oxford: Oxford University Press, 2006).

Prochaska, F., 'Voluntary action – renaissance or decline?' downloaded from http://www.historyandpolicy.org/Voluntary%20action.pdf, accessed 9 October 2006.

Putnam, R., *Bowling Alone: The Collapse and Revival of American Community* (New York: Simon and Schuster, 2000).

Rawlings B. and Yates, R., 'Introduction' in B. Rawlings and R. Yates (eds), *Therapeutic Communities for Drug Users* (London: Jessica Kingsley, 2001), pp. 9–25.

Riessman, F. and Carroll, D., *Redefining Self-Help: Policy and Practice* (San Francisco: Jossey-Bass Publishers, 1995).

Robinson, L., *Gay Men and the Left: How the Personal Got Political* (Manchester: Manchester University Press, 2007).

Robinson, D. and S. Henry, *Self Help and Health: Mutual Aid for Modern Problems* (Martin Robinson: London, 1977).

Rogers, A. and Pilgrim, D., 'Pulling down churches: accounting for the British mental health users movement', *Sociology of Health and Illness*, 13:2, (1991), pp. 129–48.

Rose, N., *Inventing Our Selves: Psychology, Power and Personhood* (Cambridge: Cambridge University Press, 1996).

Rosenthal, M., 'The therapeutic community: exploring the boundaries', *British Journal of Addiction*, 84, (1989), pp. 141–50.

Rosenthal, M., 'Therapeutic communities' in I. Belle Glass (ed.), *The International Handbook of Addiction Behaviour* (London: Routledge, 1991), pp. 258–63.

Salamon, L. and Anheier, H., *Defining the Nonprofit Sector: A Cross National Analysis* (Manchester: Manchester University Press, 1997).

Salter, B., 'Patients and doctors: reformulating the UK health policy community', *Social Science and Medicine*, 57, (2003), pp. 927–36.

Sheard, J., 'From lady bountiful to active citizen: volunteering and the voluntary sector', in J. Davis Smith, C. Rochester and R. Hedley (eds), *An Introduction to the Voluntary Sector* (London: Routledge, 1995), pp. 114–27.

Shorter, E., *A History of Psychiatry: From the Era of the Asylum to the Age of Prozac*, (New York: John Wiley & Sons Inc., 1997).

Smith, G., 'The rise of the "new consumerism" in health and medicine in Britain, c.1948–1989' in J. Burr and P. Nicholson (eds), *Researching Health Care Consumers: Critical Approaches* (Basingstoke: Palgrave, 2005), pp. 13–38.

Spear, H.B. *Heroin Addiction Care and Control: The British System, 1916–1984* (London: DrugScope, 2002).

Stacey, M., 'The health service consumer: a sociological misconception', *The Sociological Review Monograph: The Sociology of the National Health Service*, 22, (1978), pp. 194–200.

Stimson, G.V., 'British drug policies in the 1980s: a preliminary analysis and suggestions for research' in V. Berridge (ed.), *Drugs Research and Policy in Britain: A Review of the 1980s* (Aldershot: Avebury, 1990), pp. 260–81.

Stimson, G.V., 'The war on heroin: British policy and the international trade in illicit drugs' in N. Dorn and N. South (eds), *A Land Fit For Heroin? Drug Policies, Prevention and Practice* (Basingstoke: Macmillan, 1987), pp. 35–61.

Stimson, G.V., ' "Blair declares war": the unhealthy state of British drug policy', *International Journal of Drug Policy*, 11, (2000), pp. 259–64.

Szasz, T., *The Manufacture of Madness* (New York: Harper & Row, 1970).

Tantam, D., 'The anti-psychiatry movement' in G.E. Berrios and H. Freeman (eds), *150 Years of British Psychiatry, 1841–1991* (London: Royal College of Psychiatrists, 1991), pp. 333–47.

Taylor M. and Kendall, J., 'The history of the voluntary sector' in J. Kendall and M. Knapp (eds), *The Voluntary Sector in the UK* (Manchester: Manchester University Press, 1996), pp. 28–60.

Thane, P., *Foundations of the Welfare State* (Essex: Longman, 2nd edition,1996).

Thom, B., *Dealing With Drink: Alcohol And Social Policy From Treatment to Management* (London: Free Association Books, 1999).

Thompson, Mathew, *Psychological Subjects: Identity, Culture and Health in Twentieth-Century Britain* (Oxford: Oxford University Press, 2006).

Tomes, N., 'The patient as a policy factor: a historical case study of the consumer/survivor movement in mental health', *Health Affairs*, 25:3, (2006), pp. 720–29.

Tops, D., 'Stretching the limits of drug policies: an uneasy balancing act' in J. Anker, V. Asmussen, P. Kouvonen and D. Tops (eds), *Drug Users and Spaces for Legitimate Action* (Helsinki: Nordic Council for Drug and Alcohol Research, 2006), pp. 61–83.

Touraine, A., *The Voice and the Eye* (Cambridge and Paris: Cambridge University Press, Editions de la Maison des Sciences de l'Homme, 1981).

Trebach, A., *The Heroin Solution* (New Haven & London: Yale University Press, 1982).

Turner, D., 'The development of the voluntary sector: no further need for pioneers?' in J. Strang and M. Gossop (eds), *Heroin Addiction and Drug Policy: The British System* (Oxford: Oxford University Press, 1996), pp. 222–30.

Warren Holland, D., 'The development of "Concept Houses" in Great Britain and Southern Ireland, 1967–1976' in D.J. West (ed.), *Problems of Drug Abuse in Britain: Papers Presented to the Cropwood Round-Table Conference* (Cambridge: Institute of Criminology, 1978), pp. 125–35.

Webster, C., *The National Health Service: A Political History* (Oxford: Oxford University Press, 2002).

Wells, B., 'Self-help groups' in I. Belle Glass (ed.), *The International Handbook of Addiction Behaviour* (London: Routledge, 1991), pp. 254–57.

Wells, B., 'Narcotics Anonymous (NA) in Britain: the stepping up of the phenomenon' in J. Strang and M. Gossop (eds), *Heroin Addiction and the British System: Vol 2* (London: Routledge, 2005), pp. 167–74.

Williams, I., *The Alms Trade: Charities, Past, Present and Future* (London: Unwin Hyman Limited, 1989).

Wood, B., *Patient Power? The Politics of Patients' Associations in Britain and America* (Buckingham, Open University Press, 2000).

## Thesis

Efthimou-Mordaunt, A. 'Junkies in the House of the Lord', unpublished MSc dissertation, University of London, 2004.

# Index